W9-DHJ-074

Did you know . . .

the sky is really violet?

*the first passengers in space were a rodent
and a hungry feline?*

*the famous "London fog" was really
just air pollution?*

*your heart stops still for twelve years
during a normal lifetime?*

*you could count to a billion . . .
if you made it a lifetime job?*

Well, now you do!
Look inside for even *more* answers. . . .

JUST Curious ABOUT SCIENCE, Jeeves

Erin Barrett & Jack Mingo

With Illustrations by Marcos Sorensen and Spence Snyder

POCKET BOOKS

New York   London   Toronto   Sydney   Singapore

An *Original* Publication of POCKET BOOKS

POCKET BOOKS, a division of Simon & Schuster, Inc.
1230 Avenue of the Americas, New York, NY 10020

ISBN: 0-7434-2711-4

First Pocket Books trade paperback printing May 2003

10   9   8   7   6   5   4   3   2   1

POCKET and colophon are registered trademarks of
Simon & Schuster, Inc.

For information regarding special discounts for bulk purchases,
please contact Simon & Schuster Special Sales at 1-800-456-6798
or business@simonandschuster.com

Printed in the U.S.A.

ACKNOWLEDGMENTS

Special thanks to:

Amanda Ayers Barnett
Paolo Pepe
Donna O'Neill
Kathlyn McGreevy
Marcos Sorensen
Spencer Snyder
Penny Finnie
Jacquie Harrison
Steve Berkowitz
John Dollison
Anne Kinney
Jerry & Lynn Barrett
Elana Mingo
Eric Childs
Georgia Hamner
Jackson Hamner

Contents

FROM THE AUTHORS ix

OUT OF THIS WORLD! 1

NO PLACE LIKE HOME 19

WE SING THE BODY
ECLECTIC 35

WEATHER . . . OR NOT 53

TOYING WITH SCIENCE 75

PHYSICS, CHEMISTRY,
& MATH 101 87

BATHROOMS 105

GETTING AROUND 117

ZOOLOGICAL ZONE 129

MEDICINE 149

ELECTRONIC GADGETS 165

A CRITICAL MASS OF
MISCELLANY 181

THE WEEKEND SCIENTIST 199

From THE Authors

Ultimately, science seeks to answer the big questions about life, the universe, and everything. What was the beginning of time? How did the world begin? What are the forces that keep everything from falling into chaos?

We, on the other hand, have to admit that we have a special fondness for the smaller questions of science: Are smarter people's brains heavier? Why do armpits stink? Why aren't CDs and DVDs two-sided? What's a Roddenberry? What's the speed of a falling bowling ball? Why do seashells sound like the ocean?

We've thoroughly enjoyed putting together this science book hosted by the benevolent butler from the Ask Jeeves search engine. This book is the fourth in the series, following *Just Curious, Jeeves, Just Curious About History, Jeeves,* and *Just Curious About Animals and Nature, Jeeves,* and we've loved working on each and every one of them.

How did we decide what went into this book? As with the other books, we started with our river of inspiration, the secret backstage "peek box" at Ask Jeeves, where questions from users all around the world flow by in a rushing torrent of curiosity.

As you probably know, Ask Jeeves at www.ask.com is one of the most popular Web sites on the Internet, with an amazing range of information being sought day and night. True, some of them are along the lines of "Cute pictures of Christina Aguilera" or "How do I vote on *American Idol?*" which Jeeves, the digitally dynamic butler, handles with his usual finesse.

Occasionally, though, a question stands out that's clearly from a kindred soul. We don't know who these people are, exactly—perhaps someone bound and determined to win a bar bet. Possibly a younger sister trying to best her know-it-all brother. Or perhaps simply a Seeker wandering around in the night, looking for Truth by surfing the Internet. Regardless, these were the questions we picked out to answer within these pages.

For anyone who has used Ask Jeeves to uncover an offbeat answer, we'd like to express our heartfelt appreciation. Your questions have spurred us to unearth the wealth of intriguing and entertaining knowledge that we've placed in this book. Thanks to you, we now know all sorts of new stuff: that the joystick was invented in 1907, that 12 percent of all lightning fatalities happen on golf courses, that a CD groove is more than three miles long, and that 8,000 air pores in an egg allow the chick inside to breathe. So keep asking those questions on Ask Jeeves, because we're watching and reading in earnest (and in California!).

Until then, stay curious. We'll see you around the Internet.

Just Curious About Science,

Out OF THIS World!

Sure, the world's an intriguing place, but jeez, it's not as if we don't live here, like, *all* the time. Far more interesting is what started it all. Like what happened before the Big Bang. And speaking of which, ever wanted to know what astronauts really do on space missions? Grab a hand, Star Child: we're going where no man has gone before!

Fast Guys and Other Racy Topics

How fast does the space shuttle have to go to leave the atmosphere?

It has to accelerate 2,000 mph every minute for the first eight minutes of its journey to leave Earth's atmosphere. In terms of speed, from liftoff to exiting Earth's atmosphere, the shuttle goes from 0 to 17,500 mph.

Has there ever been sex in space?

Sure, there've been several experiments involving sex while in orbit. None of them, alas, have involved humans—at least, according to NASA. That makes it sound as if we don't believe them. Frankly, we do, for a couple of reasons:

1. Until recently, space trips have been relatively short— certainly not long enough for astronauts to start making "master of my space domain" wagers with each other.
2. It's against official guidelines for astronauts to engage other astronauts in sexual activity on space missions. Why would they jeopardize such a great job?
3. The astronauts stay pretty busy during a mission, with all those buttons and instrument panels.
4. Where, exactly, would one get the privacy necessary to do the deed?
5. If one guy's getting action, the jealousy among the other guys who aren't would be pretty intense.
6. Most convincing, however, is the effect zero gravity reportedly has on male genitalia (and presumably on female genitals, as well). Blood flow just isn't as efficient as it is on solid ground.

There are psychologists at NASA who study the emotional health of those in orbit. They admit that with longer and longer space missions being instigated, astronauts' sexuality should be looked at more closely, as, for most people, sex plays a vital role in staying content and happy. But officially, no specifics have been bandied about, and sex remains strictly taboo.

Oh, and just in case you were wondering whether the Russian space program is more open-minded on the issue, *Mir-25* flight engineer and cosmonaut Talgat Musabayev says, "A lot of different commissions—moral, ethical, and medical ones—that were discussing this finally ruled that one must not do it so far, because the consequences are unknown for those who would be born."

As to whether there has been sex in space up to this point, he responds, "Definitely not, although there is a lot of idle talk around this. We laugh, because we just have no opportunity for this, and the Americans are very disciplined people."

Ground Control to Major Tom

How many people have died on missions to outer space?
Have any accidentally floated away?

None have floated away. So far that's just the stuff of fiction and pop songs. As a matter of fact, all of the deaths to date have occurred during launch or reentry from outer space. The first to die was Vladimir Komarov in 1967. Several mechanical failures upon reentry into Earth's atmosphere caused his spacecraft to crash into a field.

Along with the 2003 *Columbia* space shuttle disaster, in 1971 the entire crew of the Russian *Soyuz 11* died when a valve remained open during the spacecraft's reentry. Oxygen and atmospheric pressure leaked out, and all three crew members suffocated. To avoid any similar disaster, all Russian cosmonauts wore pressure suits in subsequent missions during launch and reentry.

The next death in space was in 1986 when, soon after takeoff, the space shuttle *Challenger*'s external tank blew open in midair, killing all seven members aboard. This was the deadliest of all space accidents.

They weren't in space, true, but the accident aboard the *Apollo I* was a space program tragedy nonetheless. In 1967, during an exercise prior to countdown, a fire broke out and trapped and killed all three of the astronauts on board.

Although the *Apollo 13* astronauts made it safely back to Earth in 1970, their flight is worth noting here. In a long four-day near miss with death, the crew had to cancel their lunar landing when their craft malfunctioned. It was on this mission that flight commander Jack Swigert made the famous understated one-liner, "Houston, we've had a problem here." (These famous words got mysteriously changed along the way to "Houston, we have a problem." It's more dramatic, for sure, but not accurate.) Their flight was documented by Hollywood in the 1995 movie *Apollo 13,* starring Tom Hanks and Kevin Bacon.

> Where can I hear Jack Swigert's famous "Houston, we've had a problem" quote?

Can an astronaut land on the dark side of the moon?

No, and here's why: At the lunar equator, daytime temperatures can reach about 273 degrees Fahrenheit—a temperature that astronauts can tolerate in their space suits (although we're sure they'd prefer a slightly cooler spot). At night when it's dark, with no atmosphere to hold in the heat of the day, the temperatures plummet drastically to a chilly −243 degrees Fahrenheit, a temperature that's not as easy to withstand. This, plus a lack of visibility, makes it impractical for them to land in the dark.

But really, there isn't one "dark side" of the moon. It's dark on one side for about fourteen days, until it's rotated around and cooled on the other side for fourteen days. There is a constant *near* side of the moon as well as a constant *far* side, however. Because the moon is stuck in a steady gravitational orbit with Earth, from here we always see the same side of the moon—the near side. This is the side of the moon that all astronauts land on. Not just because it's closer, but also because if there were a big moon stuck between the spaceship and ground control, all radio contact would be lost.

Pigs . . . in . . . Spa-a-ace!

What animals have been sent into space?

The predictable list includes dogs, monkeys, rats, mice, rabbits, and fish. No pigs, as of yet, but guinea pigs and pigtailed monkeys have all been put into orbit, as well as the eye lenses of pigs, cows, and sheep.

Some not-so-likely animal crew members have included jellyfish, bees, spiders, tortoises, hornets, wasps, and incubating quail eggs. Over the years, space programs have put pretty much everything imaginable into orbit to examine the effects of high speeds and weightlessness. As a matter of fact, one of the very first launches in history was of a low-altitude, black-powder rocket in the early 1800s, which carried a feline and a squirrel a short distance into the air before parachuting back down. The rocket landed successfully with cat intact but no sign of the squirrel. It seemed one of the passengers got mighty hungry on its short journey. Needless to say, cats haven't been used much since then.

You Wanna Talk Trash?

How much man-made debris is in space?

The U.S. Space Command in Colorado Springs, Colorado, tracks about 10,000 objects in low-Earth orbit (300 to 1,200 miles up), and most, if not all, of these are man-made. They consist of about 100 space probes, 3,000 satellites (functional and not), and around 6,000 bits and pieces: lumps and chunks of debris. This sounds like a lot, but it's not even the half of it. Space Command can't detect items smaller than a baseball if they're farther than 600 miles away. Most satellites alone are at least 22,000 miles away, where objects must be considerably bigger to be seen by Space Command. One 1999 study suggested that, in reality, there were at least 110,000 objects measuring a half inch or more, weighing over 4 million pounds altogether, and traveling at speeds of 17,500 miles per hour in low-Earth orbit. Rubber sealing rings, paint, screws, whole and partial nonworking satellites, fuel tanks, and the like are just some of the trash we've put into space. When a mere flying speck of paint can dent the outside of a space-shuttle window, keeping tabs on these hurtling objects becomes a matter of utmost importance, to say the least. NASA continues to work on ways of cleaning up these hazards.

Here are some of the more intriguing items lost—or not so lost—in space:

- Edward White, astronaut on the 1965 *Gemini 4* mission, lost a glove. It stayed in orbit for a month before careening off. During its month-long orbit, it became known as the most dangerous garment in history, flying at a speed of 28,000 kilometers an hour.
- During the first ten years of its mission, space station *Mir* "dropped" over 200 items. Ironically, most of them were trash bags.
- The very first piece of man-made debris in space came in 1958 from the second satellite launched by the United States. The *Vanguard I,* although only operational for six years, is still flying around today.

Who's responsible for all the junk in space?

It Won't Work at Weight Watchers

How do I find out how much I weigh on another planet— say, Jupiter?

Multiply your weight by 2.4. That's how much you weigh on Jupiter. For the sake of clarity, let's say you weighed an even 100 pounds. On Jupiter, you'd weigh about 240. Multiply your same 100 pounds by .38 and discover that on Mercury you'd be about 38 pounds. On the moon (multiply by .17), about 17, and the sun (multiply by 27.07), 2,707 pounds. You could figure out the other planets on your own with a pretty simple equation, but you'd have to know the mass of each planet and how far its surface is from its core . . . a little over most of our heads. It's easier to search for a planetary weight calculator on-line. In the meantime, however, these four formulas are freebies. Impress your friends.

How much would I weigh on a white-dwarf star?

. . . And Bells on Our Toes

Are there rings around Uranus?

Planetary rings, found only on the outer planets, consist of pieces of ice and dust. In 1977 it was discovered that there were, indeed, rings around the planet Uranus. This was a surprise, as the rings are so dark that they'd been completely invisible up until that point. What made it possible to detect them was a star. In its orbit, Uranus passed in front of a distant star, which highlighted the planet from behind. Astronomers saw a flicker just in front of the planet as it began to pass the star, and an identical flicker as the backlit Uranus continued past the bright star. With closer scrutiny, nine rings were initially spotted around the planet. *Voyager 2,* in 1986, spotted two more.

Although Saturn's six main rings are legendary, all of the other outer planets also have them. Besides Uranus's eleven rings, Neptune has at least four, and Jupiter has a large main ring and an outer ring system, made up of many smaller rings. Saturn's rings are so brilliant, they were discovered over 400 years ago in 1659. None of the other planets' rings, however, were found until the late 1970s.

Seems a Little Loony

Is it true there's a moon crater named after Scooby Doo?

The moon has over 30 thousand billion craters that measure at least a foot wide, half a million of those with diameters over a mile. That's a lot of holes and pockmarks, and there are a lot of names that have been given to moon craters over the years. Alas, "Scooby Doo" isn't one of them. The closest in name would be the crater Scobee (over 130 feet in diameter), named in 1988 for Francis Richard Scobee, a member of the *Challenger* spacecraft crew.

There are strict and long-standing nomenclature guidelines for moon craters, and cartoon characters just don't make the cut. Besides recent additions of famous literary figures like A. A. Milne, most craters are named after historical astronomers, scientists, mathematicians, and physicists.

What you may be thinking of is the recently named rocks on Mars. One of those was officially dubbed "Scooby Doo." When the Mars probe *Sojourner* began sending back pictures of the red planet, scientists—trying to either boost public opinion of the Mars probe or prove they're really cool guys stuck in nerdy jobs—named the rocks on their screens after cartoon characters they resembled. Besides Scooby Doo, there's Yogi Bear, Casper, Asterix, Gumby, and Ratbert, to name a few. The scientists also saw animals in the boulders on Mars's surface. Names of these include Lamb, Frog, Iguana, Chimp, Kitten, Duck, and Ant Hill. Not a distinguished Ptolemy, Galileo, or Einstein to be found in this group.

> **How can I suggest a name for a moon crater?** Ask

Which planet has the most moons?

New planetary moons are discovered all the time. At this writing, Jupiter wins, hands down, with a total of thirty-nine—sixteen well-known moons and some lesser-known, more recently discovered moons. All of Jupiter's major moons are named after mythological characters. Jupiter's four large moons are Io, Europa, Ganymede, and Callisto. Jupiter's smaller moons are Metis, Adrastea, Amalthea, Thebe, Leda, Himalia, Lysithea, Elara, Ananke, Carme, Pasiphae, and Sinope.

The Sky Is Falling

I don't get it. What are the differences between comets, asteroids, and meteors?

It does get confusing, so let's see if we can clear things up a bit. Asteroids, sometimes called minor planets or planetoids, are small (in comparison to larger planets) planetlike rocks that orbit the sun. They usually reside in the Asteroid Belt, which lies between the orbits of Mars and Jupiter. Some believe they're the remnants of another planet that once resided in that spot; others believe asteroids consist of rocky debris that never quite banged together to form another planet or become part of an existing planet. Still others suspect they're the chipped-off pieces of existing planets, caught in orbit.

Comets are in a separate class of large careening space objects because they're not made of rock. The nucleus or inner core of a comet consists of frozen gases or liquids mixed with dust and dirt. Comets have been compared to large snowballs orbiting the sun. As a comet gets near the sun in its orbit, the heat predictably begins melting the outside of the frozen nucleus, forming a large surrounding field of melted gases and liquids that can sometimes trail 100 million miles behind the core. This forms the commas and tails that make comets so distinctive in the sky. As with asteroids, comets are believed by some to be the leftover parts of the icy outer planets—like Neptune, Saturn, Uranus, and Jupiter—that didn't get integrated into the planets when they were being formed.

After it's made one too many trips into the inner solar system, a comet is eventually worn down by the heat. It either becomes a fast-moving meteor, entering Earth's atmosphere, or the remaining dust and rock pieces in the nucleus become a small, orbiting asteroid.

A meteor is a "shooting star." It forms when an object—often a broken piece of an asteroid or comet—falls into Earth's atmosphere. The speeding object (called a meteoroid) hits Earth's atmosphere at about 20 miles per second. Since Earth is traveling at about 18 miles per second, the friction on impact is intense, causing the meteoroid to melt and burn, producing quite a light show. If the meteoroid doesn't completely burn up before hitting Earth's surface, it's called a meteorite. Meteorites are usually no bigger than a pebble, but can also be quite large. Arizona's famous Meteor Crater, measuring 570 feet deep by 4,180 feet wide, is an example of where a large meteorite (probably an asteroid) hit Earth about 50,000 years ago.

What's the most common element in outer space? Hydrogen.

Spin Cycle

How fast does Earth spin?

At the equator, Earth spins at about 1,070 miles per hour. As you move north or south from there, the speed slows. Exactly at the poles—Earth's axis points—the spin is much slower. Picture a record spinning. If you placed a toy at the center, then moved it an inch away from the center, then two inches, etc., the toy would have to travel farther and farther to make one full rotation. Each revolution on Earth, no matter how slow the turn, is equal to one day.

Earth also moves in an orbit. One rotation around the sun, at about 67,000 miles per hour, equals one year on Earth.

But wait, there's more. Our solar system is spinning around the Milky Way at about 558,000 miles per hour, and the Milky Way is spinning with other clusters of neighboring galaxies at the rate of 666,000 mph. Hold on tight!

What would happen if the world stopped spinning?

If Earth suddenly stopped, you'd experience quite a jolt. The atmosphere, of course, would still be in motion at the same speed that Earth had been rotating. This would lead to winds that would rub away mountains and land, smoothing Earth's surface and removing every lump, bump, nook, and cranny on it.

A gradual slowing down—as is happening now—is the far more likely scenario. Days will grow longer, leading to higher temperatures during the daytime. The Van Allen belt—the protective magnetic shield that covers Earth and protects it from intense solar winds—would presumably weaken as rotation slowed, allowing more and more dangerous solar rays into the atmosphere, mutating and then killing off every living thing. During the longer and longer nighttimes, temperatures would drop so low, no life could withstand them, either. But don't worry, this won't happen for at least another few billion years.

Do all planets spin in the same direction as Earth?

No. Uranus, Venus, and Pluto all spin in the opposite direction.

And Even More Racy Topics

How fast are other galaxies speeding away from us?

That all depends on which galaxy, as they're speeding away at different rates. On top of that, everything that's speeding away from the Milky Way is accelerating. Here's a sampling on either end: One of the slower galaxies is speeding away at about 5,040,000 mph. That's like going from one side of the United States to the opposite side in about 2 seconds, give or take a little. One of the farther and faster galaxies is going so fast, it would take it about a second to go all the way around the world. It's being clocked at about 93,600,000 mph. The general rule is that the farther the galaxy is from the Milky Way, the faster it's going.

Is Earth slowing down or speeding up?

Earth has been on a gradual slowdown since its birth. The day lengthens every 100 years by about .0015 seconds. It's not much, mind you, but in several billion years it all adds up. At that time, for instance, a month will have increased from 27.3 days to 47. But don't worry: we won't be around to experience it, anyway—the expanding sun will have gotten us by then (see page 12).

A.K.A.: I Lean

Why is Earth's axis tilted?

No one knows for sure, mind you, but many theorize that it was the result of a collision with a large planetoid during the last stages of Earth's formation over 4.5 billion years ago. The theory is that the impact with this small, rocky planet not only knocked Earth to a tilt but also tore away a chunk of Earth, forming the moon. Time and spinning smoothed Earth's and the moon's surfaces back into a sphere. Or so the theory goes. Regardless, this tilt gives Earth the seasonal variations necessary to support the variety of life—from plants to animals.

Look at the Size of Those Teeters!

What's the Chandler wobble? How is it done?

No, it's not a dance move! The Chandler wobble is a slight wobbling of Earth on its axis. It got its name from the guy who discovered it was happening in 1891–S. C. Chandler.

Still want to know how it's done? If you've ever worked on a pottery wheel, you're familiar with the concept. Earth is spinning, but it's ever-so-slightly off center in its rotation. If there were big pens sticking out of the North and South Poles that were scribbling on giant pieces of paper, we'd end up with irregular circles. In reality, these circles measure anywhere between 10 and 50 feet in diameter—not much of a wobble, really, but sometimes enough to throw you off by about one-fifth of a mile if you're navigating by the stars.

Where can I see a picture of Earth's orientation at the wobbling poles?

Size Matters

What's the biggest planet in the solar system?

Jupiter. Pluto, still officially called a planet, is the smallest.

How can the sun look so huge at sunset, and yet small at other times?

It's a matter of comparison. At sunset the sun is usually positioned near Earth's horizon instead of directly up in the sky. The mind unconsciously compares its size to the size of the objects nearby. Next to trees, houses, roads, or buildings, it looks huge. Next to blue sky and clouds, it looks smaller. Believe it or not, the sun is actually over 3,700 miles closer to

Earth at noon than it is when it's setting. When compared to the total distance that separates Earth from the sun, however, 3,700 miles just isn't that big of a deal.

Fade to Black

What's the difference between a solar and a lunar eclipse?

A solar eclipse happens when the moon's shadow crosses Earth. A lunar eclipse occurs when Earth's shadow crosses the moon. As Earth and the moon rotate, periodically they will fall into a perfectly straight line with the sun–it's called syzygy. If Earth is in the middle of that lineup, with the sun on one side and the moon on the other, Earth's shadow falls on the moon, causing a lunar eclipse. If the moon is in the middle of the line, with Earth on one side and the sun on the other, the moon's shadow falls on Earth, causing a solar eclipse. Think of the sun as a huge backlight. Which type of eclipse occurs depends on whether Earth or the moon is in front of the light. They're both amazing to see. Whereas special eye protection is needed to watch a solar eclipse, a lunar eclipse is perfectly safe to watch.

When will the next solar eclipse take place? Ask

It'll Be a Cold Day in Hell

Will the sun ever go out? What would happen to Earth?

The sun will eventually go out. By looking at the rate at which the sun is consuming its fuel, scientists have estimated that it has about 5 or 6 billion more years of life in it before it begins to fade away. Actually, it will be less of a fade than a violent burst of explosions that will end in fizzle.

As the sun's hydrogen (the sun's fuel) levels in its inner core get low, it will begin to burn hydrogen located *outside* its core, causing a burst of fiery expansion. Its parameters will expand outward, sort of like a big balloon. The balloon will extend past Mercury's and Venus's orbit and perhaps into Earth's atmosphere, killing everything, regardless of how far it reaches. The seas will boil, and life on Earth will be totally consumed by heat. Even if life survived this phase of death–the sun's red-giant stage–the ensuing white-dwarf stage would do it in. The white-dwarf stage begins this way: As the outside hydrogen is consumed, helium burns in the core of the sun and produces solid carbon. This carbon hardens and thickens into a small, glowing planet-size ball. The gases that this burning carbon core give out to the universe are collectively called a nebula. Eventually, everything but a cold black ball will be burned away, and the heat will slowly subside. The sun will enter its black-dwarf stage– pretty much the end of the line.

Does the sun make a sound?

A Nice Dinner, A Little Wine, Some Dancing

What happened before the Big Bang?
The theory of the Big Bang is that nothing–not space, not time–was here prior to the creation of the universe. About 15 billion years ago, a primeval atom exploded, sending debris off in every direction at astronomical speeds. The theory itself will never be proved. The evidence for it, however, is compelling. Solar systems are created in similar ways. We know, because we've seen them form. In addition, all galaxies are speeding away from us very quickly, still being propelled from a central source. There's also cosmic radiation throughout the universe–a common byproduct of a large cosmic explosion. None of this proves the theory, mind you. It simply makes the most sense.

What was here before our solar system?
There was a cloud of gases, the leftover stuff of dead stars. Debris–bits and pieces of various elements from rocky sources–swirled in the gassy cloud. Over time, this large cloud collapsed in on itself and compressed into a rather large

spinning disk. The disk continued to spin, further condensing the middle into a big ball. This innermost part—the bump in the center of this disk of condensed gases—grew very hot as the gases inside reacted with each other. Over time, the heat intensified, causing more and more pressure to build. Like a pressure cooker, the center of the disk blew open when the pressure inside became too great to be contained. The hot winds from the explosion blew out the debris that formed the planets—including the carbon that eventually became life on Earth. The central burning-hot core that remained became the sun. All in all, this process took one hundred million years. That may sound like a long time, but our solar system, to date, has been around a lot longer than a mere hundred million. The sun, Earth, and other eight planets are over 4.5 billion years old.

In the last twenty years, scientists have spotted similar spinning disks of gaseous clouds around new stars, confirming at least the basis of this theory.

How is a star different from a planet?

It's only a matter of size. The reason the sun "burns" is because it's so big. The weight of the sun's mass on itself crushes everything within its core—including tiny, usually uncrushable atoms. When atoms are broken, their innermost parts—the nuclei—freely bounce around and run into each other, causing nuclear reactions. Nuclear reactions produce energy and, consequently, heat.

Earth is large enough to produce enough energy to melt rock, but not enough to squash atoms. Jupiter, at eleven times the size of Earth, is large enough to produce enough pressure on its core to almost glow, but not quite. The very smallest star is still about forty to fifty times bigger than Jupiter. So somewhere between the size of Jupiter and the smallest star, mass gets heavy enough to cause nuclear reactions and glow, and technically be considered a star.

How big is the sun in relation to the planets?

Evidence of Things Not Seen

What exactly is a black hole?

A black hole occurs when a large star (much larger than our sun) runs out of fuel. The remaining mass in the star collapses in on itself, causing a vacuum and sucking everything around it—including light—into the "hole." That said, scientists have never actually seen a black hole. They are what's called dark matter—they aren't visible. So how do we know they exist? From the effects they have on visible matter around them. For instance, sometimes a large supply of magnesium, silicon, oxygen, and sulfur is detected on and around a living star. None of these gases are produced by visible stars in large amounts; they are, however, produced in large quantities when a dying star—like a supernova—is burning much, much hotter. Black holes often pull at neighboring stars, too, causing them to wobble. Scientists also know that when black holes suck in parts of a nearby star, X rays are emitted, so when big surges of X rays are detected around living stars—particularly when one or more of these other indicators is present—it's a fairly clear sign that a black hole is present.

Great Balls of Fire

I can't find the constellations in the sky. None of them look like the names they were given. How come?

It's true. For most of us, telling the difference between Orion and Gemini would be next to impossible. It would seem like the ancients—those who named the star clusters—had a little more imagination than they should have. Like cloud-watching, stargazing is a very personal thing. Each culture saw something different in the heavens. For instance, what we call the Big Dipper was to the ancient Greeks a part of their constellation Ursa Major, or "Big Bear." The ancient French and Irish called it "Chariot," while the old British dubbed it "Plough." The Laplanders? They saw this same group of stars as a reindeer. The Arabs saw our Little Dipper as a coffin, while it was a little bear to the Greeks and a spike to the ancient Scandinavians.

Interpretation aside, there is another reason why constellation spotting is hard today. Back when most of these bodies were named, the people used them as navigational tools. It made sense to find obvious pictures in the sky, which could help travelers at night find their way. What they didn't bank on, though, was that the names they chose would linger on without change all these years later, even as the sky itself was ever changing. The universe is expanding, which means that all the stars are moving and changing positions, obscuring the earlier pictures. Since stars die and come into being with some regularity, this too contributes to the changing scenery in the heavens. For instance, the main stars in the constellation Cassiopeia used to form an obvious *W;* now they look more like a squiggly line. The Big Dipper once looked more like a revolver. In about 50,000 more years, it will look like a digital number five. It's a good thing, then, that we rely on technology instead of the old star system to navigate today. If you're hell-bent on stargazing, though, astronomers recommend a comprehensive book on the constellations with exact pictures of what the constellations look like in today's night sky. They also suggest, for those more creatively inclined, that you find and create your own names for the pictures in the stars.

What's the name of the star that's closest to our sun?
Proxima Centauri, a.k.a. Alpha Centauri C. It's 4.2 light-years away.

What's in a Name?

What does the name of the star Betelgeuse mean?
Betelgeuse is the bright red star in the Orion constellation. The name is Arabic, and it generally means "the armpit of the mighty one." If you look at the constellation, you can see it firmly planted where the hunter's pit would be.

Why do some stars have normal-sounding names while others have totally uncreative names, like the "HD 209458" star recently in the news? What's up with the naming system?

Some of the brighter, more prominent stars in the sky were given formal names—usually by the ancient Arabians—while others were named by European astronomers in the seventeenth and eighteenth centuries, based on the Greek constellations they resided in. The vast number of stars in the sky, however, soon ran these old naming systems dry. Catalog systems for naming stars have become increasingly popular, especially now that we're capable of seeing so many more stars with better and stronger telescopes. Some of the catalog systems include the Bonner Durchmusterung (BD) system, the Smithsonian Astrophysical Observatory (SAO) system, the Positions and Proper Motions (PPM) system, and the Bright Star Catalog (Harvard Revised Photometry—HR), to name a few. The HD system (Henry Draper catalog) was used in naming the star you heard about in the news. Sometimes, though, especially for the brighter stars in the sky, the systems have overlapped. A good example of this is Betelgeuse. As mentioned above, it's a very prominent star in the constellation Orion, so it's been named again and again by various astronomers. Its many names include Betelgeuse, PPM 149643, SAO 113271, HD 39801, and Alpha Orionis. The surefire way to pin down a specific star, then, is to know as precisely as possible where it's located in the sky when you look for it in one of the many catalogs.

No Place
Like Home

Dust and toast, telephones and smoke detectors— for scientific curiosity, there's nothing like hanging around the house.

Stardust in Your Eyes

Is household dust the same as the cosmic dust that falls from outer space?

It's a plausible theory, on the face of it. Dust does seem like some sort of weird alien thing, in that it appears from nowhere and doesn't seem to be the same dirt, sand, and other stuff you find blowing around outside. Furthermore, it's a known fact that about a thousand tons of cosmic dust fall on Earth every day, the residue of comets and other space entities. So its easy to think that maybe a thousand tons would be enough to account for all the dust bunnies around the world. You'd be wrong, but at least it sounds more likely than some theories we've heard.

The truth is that your household dust is an individualized thing. Your dust is not exactly the same as the dust of your neighbors—in fact, your house dust can even vary in composition from room to room.

Laboratory analysis finds that dust consists of a mixture of things. For example, up to 70 percent of all house dust comes from the tens of thousands of skin cells we each shed every minute of every day. There are zillions of tiny mites eating all that loose skin, and their excrement and corpses also make up some of your dust. The good news is that the mites do a pretty good job of cleaning up most of the dead skin; the bad news is that they're the cause of almost all of the allergens that make some people sneeze and wheeze when exposed to dust.

In heavily trafficked public buildings, bits of shoe leather make up a large proportion of the dust. Other stuff in your household dust might include topsoil, industrial pollution, pillow feathers, lint from clothes, pollen, yeast, mold spores . . . and yes, probably even a little bit of cosmic dust. Libraries run by the National Trust in England have been plagued by a corrosive dust that's slowly eating away the bindings of books. The dust is made up largely of natural fibers shed by visitors' jackets and sweaters. "Visitors to stately homes tend to wear their outdoor clothes. Upper garments are far more prone to shedding fiber than lower ones, as they are more likely to be made of soft material such as wool," said a National Trust representative.

Dust to Dust

Are dust mites a type of flea?

Not really. They belong to the arachnid family, along with spiders, ticks, and harvesters (also known as "daddy longlegs").

Am I going nuts, or does my computer get dusty faster than my bookshelves?

It's not just an illusion—electronic equipment does indeed get dusty faster than other household furnishings. Dust gets drawn to electric and magnetic fields given off by TVs, stereos, and computers. In fact, it's one of the reasons so many computers are beige: Apple Computers ran tests when designing the first Macintosh in 1977 and determined that beige was the best color for hiding dust, and other computer makers followed suit.

What is the clinical name for a fear of dust?

Koniophobia. Not to be confused with *amathophobia,* which is the fear of sand.

For Every Cat Going Out, an Opposite and Equal Cat Coming In

Who invented the cat door?

Cat paraphernalia historians say that Sir Isaac Newton got tired of his cat Spithead interrupting his thoughts with all that "Let me out / Let me in!" caterwauling. So, no doubt using various advanced principles of gravity and inertia, Newton designed the first cat door.

A Toast to Dr. Maillard!

What's the "Maillard reaction"? My smart-mouthed roommate always talks about it at breakfast.

Oh, those chemistry majors. The "Maillard reaction" refers to the chemical changes that occur when you make toast. The name honors L. C. Maillard, the French chemist who in 1912 first discovered that bread's starches and sugars caramelize into intense new flavors when singed.

Pedicurean Delight

Who discovered that yeast makes breads rise?

If you've ever made bread dough by hand, you can imagine what a strain it would be to knead dough with your arms all day. About 6,000 years ago an unknown Egyptian baker took a tip from wine makers and tried using his feet instead of his hands to do the hard work. It worked really well. Legs are stronger than arms, and he got the added benefit of gravity in massaging the dough.

The real surprise, though, was when he took the bread out of the oven. His foot-kneaded dough ended up fluffy and chewy, not crisp and crunchy. The accidental byproduct of using his feet was that the bread dough got a good dose of the yeast that grows naturally between people's toes.

The practice quickly spread to other bakers. True, they didn't have a clue why foot-kneaded bread came out fluffy, but they did know that it was easier to make and that consumers seemed to like it. Despite its downtrodden origins, this softer bread became so valued that Egyptian workers accepted it as payment at the end of the workday, making them the first "breadwinners" in history.

Hello Central, Give Me 1-800-WATSON

My grandfather has a collection of old phones. Would any of them be compatible with today's phone system?

Most of them would be. Basic phone technology hasn't changed much in the last one hundred years, so that "candlestick phone" from the 1920s will work as fine as your cordless (and better in a power failure; see below). The only thing you'd have to do is adapt the plug from the four-prong type to the mini-jack now used—a simple procedure using an adapter found in an electronics store. Oh yeah, and touch-tone dialing would be out.

Why do we say a phone is "off the hook"? Did phones ever have hooks?

Yes, phones used to have hooks. They were there to hold the earpiece. "What's an earpiece?" you ask. Remember old movies where people held a bell-shaped thing to one ear while shouting into a cup-shaped thing on a boxy phone hanging on the wall? Well, the bell-shaped thing is the earpiece. Anyway, if someone's earpiece was left dangling off the hook, either by accident or because he didn't want to get any calls, other people couldn't get through to him.

After the call, people would literally "hang up" the earpiece on the hook, giving us another anachronistic term we still use today. Although handsets have replaced earpieces and cradles have replaced hooks, our basic phone terminology has changed very slowly. Which is why we still "dial" phones in order to make them "ring" on the other end, even though phones almost never have either dials or bells anymore.

Why do some phones work in a power failure, and some don't?

That's easy. The telephone network system provides its own power system through the phone lines, so that your basic corded phone will often work even when your lights won't. It's a pretty good system, but it gets messed up when you add newfangled additions to it like cordless phones. Cordless phones require household power to keep the handset charged and run the low-power radio transmitter/receiver that plugs into your wall. Our advice, of course, is to make sure at least one of your phones is of the corded variety—it could conceivably save your life in an emergency (or at least let you order pizza until you can get your microwave working again).

Pretty Cool

What's the stuff that gets cold inside a chemical ice pack?

It's kind of like magic, if you think about it: You break something inside and shake it, and suddenly the whole pack gets icy cold. What's funny is that what's inside is pretty basic stuff. You could conceivably make one yourself—all that's in it is water and ammonium-nitrate fertilizer. Breaking the bag or tube inside allows the two ingredients to mix, creating an endothermic reaction that absorbs heat, bringing the ice pack's temperature down to about 35 °F for about fifteen minutes.

The process may seem like something from modern technology, but it isn't new. Sometime around 1550, Italians discovered that mixing water with saltpeter (potassium nitrate) was a handy way to cool bottles of liquor.

I've had it explained to me before, but still don't understand how a refrigerator works. Would you try, Jeeves?

Of course.

Let's start with the two basic principles that make refrigeration possible:

1. The first rule you know already if you've ever inflated a tire using a hand pump and noticed that the pump was surprisingly hot afterward: Gases will heat up when you compress them. Conversely, gases cool down when you release pressure and allow them to expand. You've experienced these cooling properties when water evaporates off your skin on a hot day.
2. The other rule you likely learned while eating Popsicles: When two things of different temperatures come in contact, the hotter thing cools and the cooler thing heats up.

The coils and tubes of a refrigerator contain a gas. In the bad old days it was Freon, the brand name for a chlorofluorocarbon that was discontinued because it was eating Earth's ozone layer. Nowadays, new refrigerators use ammonia gas.

The refrigerator motor runs a compressor that squeezes the gas, heating it up. If it simply reduced the pressure again, the gas would quickly return to room temperature. Instead, though, your refrigerator pushes the hot, compressed ammonia gas through coils on the back or bottom of your fridge, where it loses heat to the surrounding air and cools way down, so much so that the compressed gas turns into a liquid.

The high pressure also forces the liquid through a tiny valve into a series of coils inside the refrigerator. This area has little pressure because much of its gas has already been pumped into the high-compression area. As a result, the liquid immediately vaporizes back into a gas and expands with abandon, cooling down to arctic temperatures. As it flows through coils in your freezer section, the gas brings temperatures below freezing while still being cool enough to absorb heat from the main part of your refrigerator. Finally, the gas is sucked back into the compressor to begin the process all over again.

Nervous Tick

Was the sundial humanity's first accurate clock?

Not that accurate, alas. There were too many variables in sun positioning during the year to make sundials useful for more than a give-or-take-an-hour guess of the time.

The quantum leap forward in time measurement took place in December 1656, when Dutch scientist, musician, and poet Christiaan Huygen invented the world's first pendulum clock. The clock could run for about three hours with an error rate of as little as one second. This was a great improvement over earlier mechanical clocks, which could gain or lose that much every few minutes. Huygen's clock was also the first to have a second hand—before that, there hadn't been much point in having one.

Huygen's invention was pretty revolutionary. With improvements by others over the centuries, pendulum clocks remained the standard for accuracy for nearly 300 years.

Americium, Long May She Glow

Why do smoke detectors come with a radioactivity warning?

Most smoke detectors use an ion chamber that emits a steady stream of ions that get disrupted when smoke is present, setting off the alarm. Some smoke detectors use a photoelectric sensor instead that detects changes in a light beam when smoke hits it. Unfortunately, these are more expensive, and not as effective as the radioactive "ion chamber" kind.

The radioactive stuff inside is americium, a man-made element discovered in 1945 by scientists working on the atomic bomb. The scientists created the element by bombarding plutonium with neutrons and gave it a patriotic name because there was a war going on.

The americium in your smoke alarm is made out of spent plutonium from nuclear reactors. The radioactivity itself doesn't seem to be harmful to living things while in your alarm, because the radioactive rays are absorbed by the air and alarm itself. However, that changes if the americium comes out and gets handled, inhaled, or swallowed. Discarding smoke detectors into landfills might not be a great idea, since americium has a half-life of 432 years and may have health and environmental consequences down the road.

Some Like It Hot

If mercury is so dangerous, why do they put it in thermometers? Is it the only liquid that will work?

Actually, any liquid that expands in the heat and contracts in the cold would work; technically, you could use any of them inside a thermometer. The problem is finding a liquid that has a good workable range between its freezing and boiling points. Water, for example, has a freezing point of 32 °F (0 °C) and a boiling point of 212 °F (100 °C), making it unsuitable for measuring temperatures that are hotter or colder than that.

Mercury is more suitable for temperature extremes because its freezing point is −38 °F, and its boiling point is 675 °F. Its silvery color is also an advantage in that it's highly visible.

Incidentally, the man who first used mercury in thermometers was G. D. Fahrenheit (1686–1736), the man who also invented the Fahrenheit scale. Before that, alcohol was the liquid of choice. Nowadays, because of mercury's dangers, many thermometer manufacturers have gone back to alcohol, which freezes at −175 °F (−115 °C). Alcohol is not as good at higher temperatures, since it boils at 173 °F (78 °C), but for most household purposes that's hot enough.

If cold water gets added to the water heater as soon as you use some hot water, how come the water stays hot?

Elementary science, my friend. Cold water is heavier, so it sinks to the bottom. The hot water rises to the top, which is also where you'll find the outflow pipe, the one that goes to your hot water faucet. As a result, you'll always be getting the hottest water in the heater.

Surf's Up, Little Dude

What are microwaves, exactly?

Microwaves are radio waves with a very high frequency, which makes them very short (hence the "micro" in microwave). On your radio dial, you'd hear your microwave oven broadcasting at 2,500 megahertz, if your radio dial actually went that high. It doesn't—108 megahertz is the highest a standard radio picks up. (A megahertz, by the way, is a million cycles a second.) Other radio-controlled devices that you won't hear on your radio dial include garage door openers (40 megahertz), baby monitors (about 49 megahertz), remote-controlled airplanes and cars (72–75 megahertz), televisions (54–88 megahertz for channels 2–6; 174–220 megahertz for channels 7–13; 470–890 megahertz for channels 14–83), and cell phones (824–849 megahertz).

What radio frequency does a TV remote control use?

It doesn't. TV remotes shoot a beam of infrared light.

Souper-Heated Liquid

I took a cup out of the microwave, and the soup exploded into the air, making a huge mess and burning my hand and arm. What happened?

You exceeded the liquid's boiling point, and it exploded, that's what happened.

Years ago, schoolteachers used to teach that water (unless pressurized) could never get hotter than its boiling point.

That seemed true then—you put a pot on the stove, and it would bubble and boil furiously, allowing steam to escape fast enough to keep the water from getting much hotter than 100 °C or 212 °F. Microwave ovens, however, changed the rules.

In order to start boiling, water needs hot spots, impurities in the pot surface, or small pockets of trapped air to act as "seed bubbles." All of these are plentiful in metal boiling pots, but not when microwaving in glass or ceramic containers. So, while liquids will often boil in the microwave when heated, there are times when they won't, no matter how hot they get. This is especially true with soups, because often a thin layer of fat floats on top, hindering evaporation. If you're lucky, this superheated liquid may spontaneously explode in the oven. If you're not, it may not explode until you induce an instantaneous boiling explosion by moving it or putting sugar, salt, or a spoon into it. People have been severely burned on their arms, upper body, and face.

So how do you prevent this? Well, the best idea is to not overheat liquids in the first place—for example, use the timer instead of the "zap it till it's bubbling" method. Waiting a minute before removing hot liquids from the microwave isn't a bad idea either, keeping the cup far from your face in case it blows. Since kids generally have shorter arms and less real-life experience, they are more likely to be injured; teaching them microwave safety is an excellent way to prevent accidents.

Where can I see a video of superheated liquid exploding?

How long do microwaved foods stay radioactive?

Microwaves aren't radiation, they're radio waves, so microwaved foods never become radioactive in the first place. The way microwaves work is that they have a strong effect on water, twisting its molecules back and forth rapidly. As the water molecules rub back and forth against other molecules, they heat up from the friction. Luckily nearly all foods have at least a little moisture in them—otherwise they wouldn't heat up.

Metal in the Microwave

My dumb brother left a spoon in his oatmeal when he put it in the microwave. To my shocked surprise, it did absolutely nothing—no sparks, no flames, no smoking explosion. Isn't putting metal in the microwave a big dangerous no-no?

That's true enough that you'd want to avoid doing it. However, there are some loopholes. It turns out that some metal is worse than others. According to Louis Bloomfield, physics professor at the University of Virginia, "Metal left in the microwave oven during cooking will only cause trouble if (a) it is very thin, or (b) it has sharp edges or points." Why is that? He explains that microwaves push electric charges back and forth through metal. As a result, a thin metal like aluminum foil might light up like a lightbulb filament and cause a fire. A piece of metal with pointed parts–for example, a fork or a twist tie–allows electricity to accumulate in large quantities in a small place, sending small lightning bolts flying to the microwave's metal walls.

However, a thick spoon with rounded edges would tend to simply reflect the microwaves. As long as there's food there to absorb the energy, a spoon can often emerge unscathed. Still, we don't recommend it as a regular habit.

Is it true that you can wreck a microwave by running it without something inside?

It's true. The microwaves bounce around and reflect back into the magnetron, which can essentially fry it.

Why don't microwaves escape through the mesh holes in the metal?

Microwaves can pass through glass and plastic, but they can't pass through metal. Strangely, they also can't get through a hole that's significantly smaller than their wavelength, so the fine metal mesh keeps them inside the oven, while letting some light out.

Are neon lights and fluorescent lights the same thing?

More or less. Both are long glass tubes with gas inside and an electrode on each end. Electrical currents pass through the gas from one electrode to the other, exciting the electrons enough that they emit light energy.

The only significant difference between the two types of lights is the stuff that's inside, and the colors the gases give off. Fluorescent lights have a dollop of mercury inside a partial vacuum. When the power goes on, the mercury vaporizes into gas and mixes with the thin air. As the electricity excites the atoms of the mercury gas, they glow with an ultraviolet color. Unfortunately, that color is invisible to humans. To make the light visible, the light manufacturers paint a coating of phosphors on the inside of the glass tube. When excited by the ultraviolet rays, the phosphors glow white, giving visible light.

"Neon" lights–the colorful ones in store windows–don't need the phosphors on the inside of the tube because they contain gases that glow visible colors. But technically, only the reddish lights actually have neon gas in them. To get a blue color, they use a high concentration of mercury gas. For yellow, they use a sodium gas.

Where does neon come from, anyway?

There's a little bit of neon in the air we breathe. You can extract it by using a process called *adsorption.* That involves superchilling the air at temperatures below −411 °F so it becomes a liquid. When you run the supercold liquid over charcoal, the neon molecules stick to the charcoal. By raising the temperature, you can "boil off" and capture the neon. Unfortunately, there's not that much neon in the air. Manufacturers end up having to process 88,000 pounds of liquefied air to get one pound of neon.

Turnoffs and Turn-ons

Isn't it true that you can save energy by leaving fluorescent lights on all the time instead of turning them on and off?

Actually, it's a myth that may have been true decades ago, when fluorescents required a hefty jolt to get them started. Nowadays, however, starting a fluorescent light takes only a small amount of extra electricity, so you'll save electricity by turning the light off when it's not needed.

Because the resistance through gas is less than with metal filaments, gas-filled tubes don't get very hot and use about 75 percent less electricity than incandescent lightbulbs.

Downward Mobility in the Upper Glass

Is it true that windows get thicker at the bottom over years because glass is really a slow-moving liquid?

Before we refute this story, let's look at why some people have considered the idea plausible enough to make it the pernicious urban myth it has become. Their reasoning goes something like this: Solids and liquids have different molecular structures—the molecules of solids occur in regular patterns, while the molecules in liquids are bonded in patterns that are irregular and haphazard. The structure of glass is more irregular, like liquid, so some have surmised that maybe it's just a very, very thick liquid. Casting around for evidence that would fit this theory, researchers found that antique windows from ancient buildings usually had glass that was thicker at the bottom than it was at the top. Aha! they said, here's evidence that glass slowly seeps downward with time, just like a really thick syrup would.

Well, after a lot of shouting and bickering, it looks like the "glass is solid" side won this one. Yes, glass has a different molecular structure than most solids, but it doesn't necessarily follow that it's a liquid. In fact, although scientists once called glass a "supercooled liquid," most now believe it is really an "amorphous solid," which means that it solidifies without crystals forming.

Examining other ancient glass provides the evidence. For example, scientists have examined glass bottles from ancient Rome and found no sign of seeping. They looked at telescope lenses that were precision-ground centuries ago and found no changes in shape. And they've studied arrowheads from prehistoric times made of obsidian (a naturally occurring glass) and found them to be symmetrical and razor-sharp—something that presumably wouldn't be the case if glass could droop, seep, and weep over time.

Okay, so how do they explain the bottom-heavy windows on ancient buildings? Simple. In the old days, glassmakers used a technique that all but guaranteed that each pane of glass would have a varying thickness (see below). If you were putting a fragile, heavy pane of glass into a window frame, would you place the thicker edge or the thinner edge at the bottom, where all the weight of the pane would be resting? Right, the thicker side. So rather than provide evidence of glass seepage, old windows merely prove that glassmakers (called glaziers) in the past weren't fools.

How come you can see through glass if it's a solid?

Because it's not a very dense solid. There's enough space between the molecules to let light shine through.

How do they make glass, anyway? It's basically melted sand, right?

Most glass is made of silica sand (silica, or silicon dioxide), with a little soda ash (sodium carbonate) and limestone (calcium carbonate) added to lower the sand's melting temperature.

Early humans got the idea for making glass after finding chunks formed by lightning strikes on the beach. These early humans originally cast it like metal and then ground and polished it, which was a laborious process. At about 50 B.C., Romans figured out how to blow glass using hollow metal tubes, which suddenly made glass inexpensive and practical for a variety of cups, bottles, and other vessels.

Panestaking Work

How do they make glass windows so flawlessly flat?

Making nearly perfect glass long eluded human ingenuity, as witnessed by the panes in old houses, in which imperfections in the thickness and surface create a subtle "funhouse mirror" effect, distorting what you see outside.

The ancient Romans never figured out how to make sheet glass for windows. They did try molding window glass and then grinding and buffing it, but the results were very expensive and not very good. They didn't progress much further, in large part because in their climate, they really didn't need windows, anyway.

Necessity bred invention in much cooler Germany, and the glass window was invented there about A.D. 600. They'd blow a large sphere and wave it around so that it elongated into a hollow cylindrical shape. The glassblower then quickly sliced the cylinder open and flattened the glass onto a metal table. However, this glass was thick and hard to see through clearly, so Norman glassmakers came up with a clever method of making much thinner windows that was the state of the art until the late nineteenth century. Although thin, these glass sheets were invariably thicker at one end, leading to the urban myth that glass is a liquid (see above). Using about nine pounds of molten glass, the glassmaker blew a shape that looked like a huge Florence flask (a scientific beaker with a round body and straight neck on top). Then he'd attach a metal cap to the center of the round end, flatten the flask into a decanter shape, attach a metal rod called a punty to the metal cap, and remove the blow pipe, leaving a hole. Now the real fun began: the glassmaker spun the punty rod inside a flashing furnace. To quote an 1860 account by college professor Sheridan Muspratt:

> The action of heat and centrifugal force combined is soon visible. The nose of the piece, or hole caused by the removal of the blowing pipe, enlarges and the parts around cannot resist the tendency. The opening grows larger and larger; for the moment is caught in a glimpse of a circle with a double rim; the next moment, before the eyes of the astonished spectator, is whirling a thin transparent circular plate of glass. . . . The sound of the final opening of the piece has been compared to that produced by quickly expanding a wet umbrella. In this way a flat circular disc, sixty inches (five feet) in diameter is produced, of almost uniform thickness, except at the point of attachment to the punty and the glass at the edge of the disc is also in some cases a little thickened. . . . The cutting of a circle into rectangle sheets, must necessarily be attended with waste and confined to fairly small sizes.

Nowadays, however, making windows is a lot easier and more efficient. The method the glassmakers use is to float molten glass on top of a bath of molten tin. The glass hardens in a uniform thickness without coming in contact with anything solid, since tin melts at a lower temperature than glass. As a result, the process creates panes that are flawless and smooth on both sides.

We Sing THE Body Eclectic

Genius Internetus

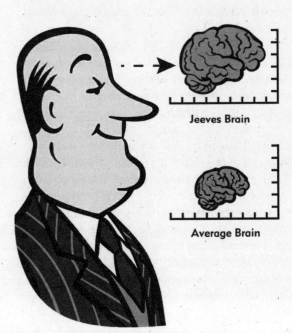

Jeeves Brain

Average Brain

Babies spend hours and hours looking at their hands, tasting their toes, and listening to the sound of their own voices. We never really outgrow that completely. It's no wonder Ask Jeeves gets so many questions about the amazing human body.

Brain, Brain Go Away

How often does the brain replenish itself with new cells?
Although recent studies have shown that the brain can grow some new cells, you'd be wise to take good care of the ones you were born with. You can't generate enough of the right kind to even consider riding your bike without a helmet, young man.

If we use only 10 percent of our brain, do we know what the other 90 percent of our brain is for?

Ah, the old "we use only 10 percent of our brains" myth, so beloved by psychics, human potential gurus, and others who have made a cottage industry of hawking seminars and books. It's not true. As it happens, you probably use all of your brain during a normal day, but not for every function. It's just like your body's muscles—just as you don't use all of your muscle groups at one time, you also don't use all of your brain at once. But don't give up on that other 90 percent—between eating, sleeping, working, and finding your keys, every bit counts.

Where can I see brains of other species?

Ask

Are smarter people's brains heavier and bigger than dumber people's brains?

No. There's no apparent connection between brain size in humans and their intelligence. If that were the case, the average man would be 10 to 12 percent smarter than the average woman because men have larger brains by about a third of a pound in weight. For that matter, the average elephant would be five times smarter than the average human.

Researchers have discovered, by subjecting men and women to a series of tests, that it's not size that matters, but the amount of available gray matter—the stuff you have to think with. Women have a little bit more of this than men, the experts say. What men have that makes their brains bigger is white matter. That's the part of the brain that deals with spatial information—the concept of knowing where you are in relation to other things. Men, apparently, have a lot more of this than women, accounting for the brain-size difference. Perhaps also accounting for why men hate to stop and ask for directions.

Only Skin Deep

What part of the body has the thickest skin?

It's pretty much as you'd suspect. The soles of the feet and palms of the hands have the thickest layer of skin. After that, the back and nape of the neck are the thickest. The thinnest layer of skin is around the eyes—particularly the eyelids.

Why do toes and fingers wrinkle in the bathtub?

Actually, they aren't wrinkling at all but puckering from swelling. The thick outside layer of skin, called the stratum corneum, takes on excess water when saturated–sort of like when dried beans are soaked in water. The underlying skin and connective tissue around the stratum corneum doesn't absorb water and therefore doesn't swell along with it. This anchoring of the skin tissue layers makes the swollen area appear puckered.

It's not just the fingers and toes that do this, though they are easier to see because the stratum corneum is thickest in those areas. When you see your fingers and toes wrinkling, take it as a sign that your entire body is on guard, preventing water, soap, dirt, and germs from invading through the skin.

Masticating with the Cavemen

If everyone has them—foolish and wise—then why are they called "wisdom" teeth?

They're called wisdom teeth because they're the latest to arrive–when a person's (presumably) older and wiser. The funny thing about our molars is that none of them form beneath any of the baby teeth. They develop only after our jaws grow big enough to make space for them. The problem arises when your jaw doesn't quite grow big enough for all the molars that want to come in.

Wisdom teeth are fascinating things–if you forget the pain and complications they cause, that is. The complications–the fact that so many people no longer have room for their wisdom teeth–allow you to see evolution in action. Early humans needed these extra molars to help them chew the tough fibers in meat and vegetation. Being closer to the jaw gave the teeth better leverage. People's jaws were larger and protruding, in order to make room for all of the teeth necessary for masticating a good meal way back in 10,000 B.C.

But ever since humans discovered how to tame that blazing red fire, we have no longer needed those teeth to survive. We've learned how to cook tough foods well enough so that they can be eaten without much effort. The size and shape of our faces and jaws have changed to deal with these lifestyle advancements. Unfortunately for many, our teeth haven't completely taken the hint: we still try to grow these extra molars in the back, despite the fact that we no longer need them and often don't have space for them.

Do other animals have wisdom teeth?

Other mammals do have what we call wisdom teeth—the molars that lie farthest back in the mouth. Often in humans, the jawbone is too small to accommodate these teeth, so they can't break through the gums and become painfully wedged against the jaw and the other molars. In other mammals, however, their jawbones are still large enough to house them. They're referred to as "third" molars, and look no different than the other molars in their mouths.

Dot Your *T's*, Cross Your Eyes

You're cross-eyed if your eyes point inward. What's it called if they point outward?

The opposite of cross-eyed is wall-eyed, but any kind of deviation from both eyeballs pointing straight ahead is called strabismus. The technical terms for these conditions are *esotropia* (cross-eyed) and *exotropia* (wandering eyes, or wall-eyed). *Hypertropia* is when one eye wanders up. About 5 percent of all children have some degree of strabismus, which most of the time can be corrected with surgery.

Has there ever been a case where someone's crossed his eyes, and they've stuck that way?

No, no matter what your older brother might tell you.

Why are Asian eyes slanted?

Asians don't actually have slanted eyes—that's an optical illusion, as it were. Folks from Asia actually have an extra fold in their eyelids called the epicanthic fold. It's at the top of the eyelid and is elongated, sometimes extending beyond the eyelashes. Caucasian or African eyes have a slight, shortened fold where the epicanthic fold would be.

Evolutionary biologists believe there's a good reason for this difference in eye structure. During the last ice age, when

Asians were trapped in the frozen tundra of Siberia, large, round eyes with thin-skinned eyelids were an obstacle to survival in that they were more likely to be damaged by the cold. Round-eyed people tended to not survive as well as those with an elongated fold of fatty tissue that insulated the eyeball and inner lids and who therefore had less direct exposure to the cold air. Another survival characteristic that they developed was flatter noses–big protruding Caucasian noses would be vulnerable to frostbite in such conditions.

How much does an eyeball weigh?

About one ounce. You could send one in the mail for the price of a U.S. postage stamp. But make sure you mark it "Fragile."

If people are born blind but later have corrective surgery, how do they come to grips with having a new sense?

Not so well, by many accounts. The medical miracle of correcting blindness works better with those who've gone blind over the years, or those who were born *almost* blind–but not completely.

There does seem to be a problem with suddenly getting sight if you've never had it before. Normally, the brain learns how to process visual information from the time you're born until about six or seven years of age. Once you're beyond these important years of brain development, it's very difficult for the brain to learn how to see and interpret these visual sensations. A person born blind will be able to understand some of what the eye can perceive, but their brain will never process the sights in the same way an always sighted person can.

I'm Dexterous, but You're Sinister

Who gets injured more often: right-handed people or southpaws?

In Europe and America, lefties make up 11 percent of the total population. And to answer your question, left-handed people are injured more often. Because they live in a world made for people who are right-handed, some recent studies have suggested that left-handed people are born with a weakened immune system and tend to have shorter life spans. Many lefties are suspicious, though, that this is just more propaganda from a right-handed world.

What's the technical term for left-handedness?

Sinistrality. Right-handedness is called dextrality. It's a sad case of right-handed bigotry that the word *sinister* comes from the name for left-handers, and *dexterity* from the name for right-handers.

Where does the word southpaw come from?

The origin of the term reaches all the way back in time to, well, baseball. All baseball diamonds are created equal: the pitcher faces west, the batter faces east. This helps the batter survive a careening pitch by keeping the setting sun out of his eyes. It also means that the pitcher's left hand will be at his south side, hence the term "southpaw." The term was coined by Charles Seymour, a Chicago sportswriter.

Have a Little Heart

If your heart stops, can you still live?

Sure, it happens all the time. The heart beats more than two and a half billion times in the average lifetime. In between each beat, the heart stops momentarily, meaning a total of twelve years of your life is spent with a stopped heart.

Other than that, hearts can stop for a few minutes and still be resuscitated, thanks to CPR and defibrillators. Although we should note that television portrays the incidences of people recovering after CPR as a lot higher than in real life. Alas, most people whose hearts have stopped won't be resuscitated. For example: while on TV, most patients get their lives saved after getting emergency CPR, the success rate in reality is remarkably low–fewer than 15 percent of those who have had a heart attack in a nonhospital setting.

Why Be a Gloomy Gus When You Can Be a Fun Gus?

Why does the tip of a penis look like a mushroom?

First, some background. The tip of the mammalian penis is called the glans. Although "glans" sounds a lot like "glands," it actually has nothing to do with the production of hormones. *Glans* comes from the Latin word meaning "acorn."

It's an odd sort of shape, isn't it? One many of us have wondered about–biologists included. One theory is that the shape has been evolutionarily successful because it helps create a mild plungerlike suction within the cervix. This may sound like a strange goal, but seeing that humans were once as promiscuous as chimpanzees, the suction increased the chances that the current mate's sperm were the ones doing the fertilizing by mechanically removing the competition (i.e., sucking the previous sperm out).

We wonder about this one, however. It would seem that any suction strong enough to remove a competitor's sperm would cause some pain and perhaps even bodily damage.

There's a much more likely reason for the shape of the penis that doesn't require such a stretch of the imagination. The glans has a large number of nerve endings–more nerves than any other single part of the male body. It's designed this way to produce pleasure. In terms of development, the larger the glans, the more nerve endings, therefore the more pleasure gained during sex. To further this as the main "reason" for the shape, the female clitoris–the main sexual pleasure center in the female body–also has a glans at the tip. The female glans clitoris likewise has 8,000 nerve fibers, about the same number as the glans penis, but in a more condensed area. So it's likely that the glans is shaped that way mostly to increase surface area and enhance pleasure. The pleasure gives motivation to do what needs to be done to keep the species going for another generation. That the glans also adds some padding at the end of a potentially painful device and that it can help to form a tight seal with the cervical opening at a crucial moment of climax are additional reproductive bonuses.

Hey Big Boy, Feeling a Little Sheepish?

Is it true that the first condom was made of sheep intestines? What other kinds of birth control have people used?

The first condom on record was made from a goat bladder by ancient Romans. But it wasn't the first form of birth control. For at least 8,000 years, folks used coitus interruptus as a widespread and moderately effective form of birth control. It's mentioned in both the Talmud and the Book of Genesis as a way to prevent pregnancy.

But that's a tame and relatively benign form of birth control compared to other early attempts. For example, people used barriers that often contained acidity compounds as a spermicide. This included slices of lemon, mixtures of lint and acacia tree powder, dried crocodile dung, figs, beeswax, and mustard seed. Antiquated oral contraceptives were pretty dangerous. Although most people opted for mixtures of herbs and roots, some ingested such substances as gunpowder, quicksilver, arsenic, and even camel spit in the belief that they would prevent pregnancy. Thank heaven for modern medicine.

No Milk Today

Why do men have nipples at all, Jeeves? I mean, what good could possibly come from them?

No good at all, dear reader. None at all.

It's just more efficient that way. Nipples are one of the body parts that everybody gets issued early in life. They're formed during the first fourteen weeks of development in the womb. The fact that everyone gets them sort of makes sense, since they begin growing before male or female hormones start flooding the fetus and instigating the development of sexual organs and other gender-based characteristics.

Nipples are interesting things. While some argue that nipples and hair are what make mammals distinctively mammalian, that's not entirely true. For starters, there are exceptions in the mammalian world, like the monotremes—the duck-billed platypus and the spiny anteater of Australia—that lack mammary glands and nipples in both the males and the females. Instead of the usual arrangement, the mothers excrete milk through pores in the skin, and the young lap it off their fur.

Secondly, there are a few male mammals that don't have nipples—the stallion and the bull, for example. This may be because the mammary glands of the females of these species exist back between the hind legs. In the males, there's simply no room to develop nipples there—they have other, more important equipment that needs the space in that spot.

The bottom line is that nipples on males serve no apparent purpose. However, the romantic within us wants to believe that they're there to remind us that the genders are not so different after all. Group hug!

The Nerve of Some People

What's it called when the sun makes you sneeze?

It's called a photic sneeze reflex, and it occurs in about a quarter of the population.

What happens to convince your nose that the sun is an object that must be sneezed out of your nasal cavities? It's simply a case of crossed wires, like hearing interference on your phone line from someone else's conversation. There are so many nerves running through your facial and nasal cavities that some of them end up right next to one another. Stepping into bright light sends rapid signals through your optical nerves to your brain so that it can instruct your pupils to dilate and your lids to squint in response. Some of these optic nerves butt up against the nasal nerves and trigger a wildly inappropriate response—a sneeze.

Do you cough or choke while cleaning your ears? A similar signal-crossing scenario may be taking place.

Why do fingernails scratching across a chalkboard send me through the roof?

After serious research into the matter, it's now understood that it's actually some of the lower-frequency sound vibrations involved in this action that trigger the aversion response, and not the squeaky higher ones you might suspect.

But why do some people have this response? Some researchers say it might be a physiological throwback to the wild jungle days of human history. As evidence, they point to the fact that the screeching chalkboard sound resembles the danger sounds of some primates. Based on our simian heritage, humans likely had a similar warning sound in the distant past, and it may be that when we respond to a chalkboard sound, we're hearing the residual echoes of a warning hardwired into our systems a long time ago.

Researchers have also found that the sound of Styrofoam being rubbed together is the second most irritating sound next to fingernails on a chalkboard.

Questions from the Ten-Year-Old Boy Within

I've heard that eating asparagus makes your pee stink. I eat it all the time, and I've never smelled anything. What gives?

People have been fascinated by this phenomenon for eons. Why, even Ben Franklin had something to say on the subject. He wrote, "A few stems of asparagus eaten shall give our urine a disagreeable odor; and a pill of turpentine no bigger than a pea shall bestow upon it the pleasing smell of violets." Turpentine? Well, anyway, back to asparagus. As we were saying, it's been a curious thing for quite some time. A researcher in the late nineteenth century actually thought he'd identified the substance that causes the stink: methanethiol. He obtained these results by getting four men to agree to offer urine samples, then eat three pounds of asparagus each. After fifteen minutes or so, he had them urinate again, and when he compared the samples, he found that each of them suddenly contained larger quantities of methanethiol. This was the standard explanation for years, until about 1975, when yet another researcher conducted similar tests. His results were a little different. He found that the odor derived from the compound S-methyl thioesters.

Neither explanation has been ruled out, and most sources list both possibilities instead of taking sides. Let's just say that either one of these results can be summed up by stating that the stink occurs when the body produces sulfur compounds during the digestion of asparagus.

But your question is about why your urine doesn't smell after eating asparagus. Well, believe it or not, this too has been tested. Only about 22 percent of those polled claim to have smelly urine after eating asparagus. So why them, and not the rest of us? Well, when those same 22 percent were asked to smell the urine of the other 78 percent, guess what? They could smell the sulfur compounds at work. The same was true in reverse: those 78 percent who claimed not to have stinky urine couldn't smell a thing when asked about the smell of the other 22 percent.

So the answer is that your urine does smell when you've eaten asparagus, you just can't smell it.

In a situation where you don't have any other disinfectant, like on the show Survivor, is it really better to use pee than nothing?

It can be, yes. Perhaps you've noticed this when changing your cat box, but one of the main ingredients in urine is ammonia. This means that urine can act as a disinfectant. Ammonia is also used to take out the stinging and itching of insect, jellyfish, and anemone encounters. In our germ-conscious society, you may not be aware that cultures throughout history have in fact used urine for medicinal purposes. For some time, urine was the main ingredient in toothpaste. The astringent properties helped kill plaque-causing bacteria in the mouth. But don't try this at home; the stuff in modern toothpaste and Listerine works even better against our starch- and sugar-rich diets.

Why is my pee different shades of yellow at different times of the day?

A higher ratio of wastes to water turns urine a bright yellow color, which explains why the first urine in the morning is often quite a deep yellow. If your urine is a very faint yellow, it generally means you're well hydrated.

Why do beans make you break wind?

Blame it on the oligosaccharides, or, if you prefer, the dog. Oligosaccharides are a special sugar found in beans and other fibrous foods like soybeans, cabbage, peas, and cauliflower. These sugar molecules are very large and are not digested by the small intestines, so they pass on to the large intestines, where colonies of bacteria lie in wait. These bacteria latch onto the large sugar molecules and multiply rapidly, releasing gases as they go. With no place to go but down, the gases invariably reach your colon the moment you step into an elevator or meet a blind date.

By the way, dog food contains high quantities of soybeans, which is why our pooches make a perfect excuse for our own little windy accidents. However, a pet cow would make an even better scapegoat. Because of its grassy diet, one dairy cow can pass 312 pounds of methane gas in just one year.

How flammable are farts?

Pretty flammable. Of the three main gases in farts–nitrogen, methane, and hydrogen–methane and hydrogen are both highly flammable. Nitrogen's not. According to the statistics we found, over a quarter of the idiots who try lighting farts get burned. The rest are treated to a barely visible show of yellow and blue flames, as well as a putrid odor.

Are burps flammable, too?

Burps are produced when you swallow air. Normally, this can happen while eating and drinking, but it's possible to build up air in the stomach through the regular, automatic swallowing that your body does to clear away draining saliva and mucus. The bubbles in carbonated drinks will often cause burps as well. But the digestive process that produces the smelly and flammable compounds in farts takes place in the long slow trip through the intestines, not in the stomach.

Since air and carbon dioxide are not flammable, you can burp near an open flame without fear of conflagration.

Water, Water Everywhere

How many glasses of water do I need to drink every day to get enough liquids?

For years we've heard there's a fixed quantity of water we should drink every day. It turns out it was just a marketing ploy by companies that sell bottled water. We guess it goes to show that if you can't trust people who'll sell you a plentiful, virtually free substance for a buck-fifty a bottle, you just can't trust *anyone* these days.

If you think about it, it does seem a little suspicious. Most of us live in places where the water from the faucet is pure, tastes good, and is fortified with fluoride. So why would we need to purchase our water in bottles from some company?

Not surprisingly, this was what the water bottling companies were asking themselves, too. In response, they came up with a great way to keep people buying their product by claiming that every person needed to force down at least eight glasses (or six conveniently sized, easy-tote bottles) every day.

It was brilliant marketing in that it sounded reasonable enough. People bought the theory so thoroughly that most of us now intrinsically believe there is some fixed dose of water we need to guzzle daily to stay healthy.

It turns out, though, that all we really need is to make sure we drink when we're thirsty—the exact advice our mothers, our bodies, and our common sense have been giving us for years. It's not that bottles of water aren't convenient, but tap water will do the same thing, besides being cheaper, readily available, and fluoridated for healthy teeth. Furthermore, you know the source—something you can't always bank on with the bottled variety.

With a Friend Like This, Who Needs Enemas?

My girlfriend's trying to talk me into getting a high colonic. But why would you need to clean your colon? Help me out here, Jeeves.

First, a little explanation for those more refined readers among us: a "high-colonic irrigation" is a high-powered enema in which up to twenty gallons of liquid are pumped into your intestines a few pints at a time, with a really smelly liquid pouring out afterward. A few "alternative health" gurus claim that cleaning out your intestines in this way will lead to a wide variety of benefits that is (literally) unbelievable.

Exactly what are the benefits of having a sparkling clean set of intestines? Well, how about this list, from the Web page of a high-colonic provider: "By cleansing the colon of impacted and putrefied waste matter on a regular basis, high colonic irrigation offers relief from a variety of disturbances such as: chronic fatigue, excessive gas, headaches, irritability, depression, skin problems (psoriases), lethargy, constipation, chronic diarrhea, parasites, yeast infections, colds and flu, distended abdomen, bad breath, insomnia, allergies, etc."

Whew! How could anyone argue against that? Well, according to medical boards across the country, the procedure of ramming water up your intestines is potentially dangerous. Hazards include death and illness from electrolyte depletion, contamination by the colonics equipment, and even the perforation of your intestinal wall. Considering that the medical community considers this procedure completely worthless from a medical standpoint, it doesn't seem to be worth the risk.

If you need a backup argument for your girlfriend, start with the human body's amazing ability to clean itself. No matter what she's heard, it is not true that red meat sticks in your intestines, or that gum gets lodged there for five years, or that you can be starved for nutrients because of a buildup of "toxins and fecal matter." Like the skin, the colon sheds its lining every seven days or so, so stay out of its way and let it do its job.

A Pore Odor Indeed

Why do my armpits stink?

Because bacteria love dark, wet places. You sweat there, creating moistness, and because your arm isn't always raised–which would allow constant airflow–bacteria begin to multiply. You don't really stink, per se–it's the excrement of the bacteria that stinks.

How does deodorant work? Is it anything other than a perfume that hides the smell?

The only thing at work in a deodorant is fragrance. It may be natural fragrances from oils, or it may be chemical perfumes, but when you use it, all you're doing is masking the smell. An antiperspirant, however, contains aluminum or zirconium to block pores so sweat can't escape from the sweat glands as easily, decreasing both wetness and the reproduction of those smelly bacteria.

Do sweat and body oils come from the same pores, or are they totally different?

They're totally different. Sweat glands produce sweat, and sebaceous glands produce oils. Most of your sebaceous glands are located on your scalp and face. When something obstructs the regular release of sebum (fatty oil), it leads to painful and blocked glands, also known as acne.

Oily to Bed, Oily to Rise

Are hair oils the same as body oils?

Yes, both come from sebaceous glands on the scalp. Unfortunately, we end up washing most of them away before they have time to work their magic on our dried, cracked hair . . . but hey, that's why we have conditioner.

What do conditioners do besides detangling hair?

Not much, but here's what they're doing. Have you ever looked at a strand of healthy hair under the microscope? It's crackly, with a lot of crevices up and down the length of it. Conditioner's primary function is to fill in those cracks and coat the hair, allowing it to slide smoothly against your hairbrush and your other hairs, reducing tangles and protecting against damage. The oil in the conditioner makes your hair shinier, too.

Them Bones Gonna Walk Around

Are double-jointed joints really different from regular joints?

No. As a matter of fact, no joints are truly "double-jointed." Instead, the joints and connecting muscles are merely more flexible and mobile, bending farther back than most people are capable of.

Although some people are born with a greater range of mobility in some or all of their joints, practice can produce similar effects. Contortionists and gymnasts, for example, often stretch the joints backward to achieve greater flexibility.

What exactly is the funny bone?

What we call a "funny bone" isn't funny and it isn't a bone; it's a nerve called the ulnar nerve, which is entirely too exposed for its own good and yours. When it is hit, tingling, numbness, and pain strike the outer fingers on your hand and may take seconds or even minutes to subside.

Why is it called the funny bone?

The name comes from a wordplay on the name of the bone that runs from your shoulder to your elbow, the "humerus." Now that's funny.

Greetings from the Interior

Why does my stomach growl?

It can be embarrassing when stomachs make noises, but they have their reasons; a lot is going on in there. When your stomach is empty, and there's nothing for the stomach acid to do, it doesn't just go away. It churns and produces gases as it bubbles and waits, impatient for you to feed it. When food does come down the pipe and drop into the brine, the digestive juices make noises as they work on the food as well.

I've heard so many different theories, Jeeves. What is the best way to cure the hiccups?

Hiccups are muscle spasms in the diaphragm and throat. The large muscle at the bottom of your chest cavity sometimes spasms (as do most all of the muscles in the body, at one time or another). When this happens, air gets sucked in through the mouth. The air never actually reaches the lungs because another spasm occurs in the muscles of the throat at the same time. Any number of remedies are suggested to stop hiccups. Some of the more common ones include breathing into a paper bag, drinking a big glass of water (sometimes from the opposite side of the glass—or is that just a practical joke?), and massaging your belly. But basically the same principle works here as with other muscle spasms: If you relax, calm down, take your mind off of the hiccups, and get on with your activities, the diaphragm will stop spasming all on its own.

Having said that, though, here is the only cure for hiccups we know of that actually works time after time. We call it the Mary Poppins Cure, after the song from the movie: take a "spoonful of sugar" and swallow it dry. Your hiccups will be gone in seconds.

What are the lyrics to the Disney song "A Spoonful of Sugar (Helps the Medicine Go Down)"?

Weather . . .
OR Not

From stormy weather to raindrops falling on your head, from a foggy day in London town to here comes the sun, everybody talks about the weather . . . and Jeeves is here to answer your questions about it.

And a 30 Percent Chance of Meteor Showers

Why is the study of weather called meteorology? What is the study of meteors called, then?

Let's blame Aristotle. He was a smart guy who was right about a lot of things. However, he was wrong about meteors. He coined the term *meteorology* in 340 B.C. in a book called *Meteorologica* in which he laid out his ideas of weather. He believed that meteors were part of Earth's weather system, that they were pockets of gas that evaporated off Earth during the day. When they reached the lowest celestial sphere that surrounds Earth, he explained, they sometimes burst into flame.

Aristotle's theory was accepted by the scientific establishment for more than two millennia, until the 1800s, when astronomers took over the study of meteors, rightly identifying them as space particles. We still use the confusing names, even though we know that meteors and meteorology are not really related.

Or are they? The term *meteor* comes from the Latin *meteorum,* meaning "atmospheric phenomenon." Aristotle liked the term so much, he used it to describe a number of weather occurrences. He called winds "aerial meteors," rainbows "luminous meteors," rain or snow "aqueous meteors," and lightning or shooting stars "fiery meteors."

Air Apparent

How long has smog been a problem in cities?

Air pollution was a problem long before cars and trucks—in fact, long before the Industrial Revolution. No doubt the tens of thousands of fires used for cooking and heating often created lung-unfriendly environments in ancient cities.

Wood smoke was bad enough, but things got worse as humans discovered other sources of heat. For example, London grew so large during the 1200s that its shrinking forests couldn't supply enough fuel. As wood prices soared, Londoners began burning "sea-coal" from off the northeast coast to heat their homes and fuel their factories.

The soft, bituminous coal was cheap, but it produced a great deal of smoke when it burned—so much so that it actually changed the weather patterns in London, creating a near-permanent state of smoky "London fog" (it wouldn't be called smog until the early twentieth century). The smoke particles in the London air condensed, forming a fog of tiny chemical-laden water droplets. One of the chief components of the fog was sulfur dioxide, which attacks the lungs, making breathing difficult.

Despite attempts to deal with the problem (see below), the coal-befouled air was the hallmark of the great city for centuries. Shakespeare's witches in *Macbeth* even pay tribute to the city's air: "Fair is foul, and foul is fair: Hover through the fog and filthy air."

In 1873, one especially bad "fog" caused 268 deaths from bronchitis. In 1879 the city had a sunless winter as the smoke wrapped the city in an unremittingly gloomy darkness from November to March. In 1902, a fog monitor wrote that the coal-tar-drenched air made it difficult to see across the street some days: "White and damp in the early morning, it became smoky later, the particles coated with soot being dry and pungent to inhale. There was a complete block of street traffic at some crossings. Omnibuses were abandoned, and several goods trains were taken off."

About the same time, a French scientist attending a health symposium in London coined a new name for London's affliction. On July 3, 1905, Dr. Harold Antoine Des Voeux blended the words *smoke* and *fog* and came up with "smog."

Was the Clean Air Act in 1970 the first law that tried to reduce air pollution?

Not even close. In 1272 King Edward I banned the burning of the notoriously smoky sea-coal under penalty of torture or death. Even after the king had the first offender put to death publicly, the law deterred no one. Few people could afford wood, so out of necessity they continued to burn sea-coal, and the air quality didn't improve. In fact, the pollution levels only worsened. Both Richard III (1377–1399) and Henry V (1413–1422) also tried to curb the use of sea-coal, to little effect.

Subsequent kings didn't even bother trying. In 1661 author John Evelyn wrote an anticoal treatise called *FUMIFUNGIUM: or the Inconvenience of the Aer and Smoake of London Dissipated* that called on the authorities to do something about the problem. "And what is all this, but that Hellish and dismall Cloud of SEACOALE? So universally mixed with the otherwise wholesome and excellent Aer, that her Inhabitants breathe nothing but an impure and thick Mist accompanied with a fuliginous and filthy vapour."

Despite some killer fogs during the 1700s and 1800s that caused smog-related fatalities, not much changed in the world's most polluted city until the middle of the twentieth century. Up to that point, since coal fueled the Industrial Revolution, people who complained about the air were accused of being against "progress." However, in 1952—with car and diesel fumes added to the smoggy mix—an especially bad December resulted in at least 4,000 human deaths (as well as deaths of entire herds of cattle in the fields outside London). The British Parliament finally agreed to pass its own Clean Air Act in 1956, reducing coal-burning, establishing "smokeless" areas of the city, and moving power stations to places outside the city limits.

Pigment of Your Imagination

Air isn't blue, so why is the sky blue? And why is a sunset red, for that matter?

The sun emits light waves in all colors of the spectrum, with the wavelengths of each color vibrating at different rates. The reds and yellows have a long wavelength. Since they're traveling in straighter lines from the sun to Earth, they are covering more ground with less chance of interference. As a result, fewer of them get sidetracked by air particles, dust, and other stray molecules as they pass through Earth's atmosphere.

The blues and violets, on the other hand, have the shortest wavelengths and vibrate very quickly. This makes them more likely to collide with the various particles in the atmosphere and bounce and scatter wildly in random directions.

What we see in a blue sky is an abundance of these scattered blue and violet colors. Although there is more violet than blue light in the mix, the blue is more visible to our eyes, so the sky looks blue to us.

At sunset, the sun's light gets angled through more of Earth's atmosphere than when it's directly overhead. (If you have trouble picturing this, pretend the inner part of an orange is Earth, and the peel is its atmosphere. If you poke a pin straight through the peel, you might reach the orange in a sixteenth of an inch. But if you angle the pin as you stick it in, you might go half an inch before you pierce the skin.)

Simply by adding more atmosphere, you've already added more air molecules, more dust, and numerous other random particles. Air pollution adds even more particles for the blues and violets to bounce off. As a result, most of the blues and violets get bounced away from our eyes, making sunsets look more brilliantly yellow and red.

Sunrises travel through the same amount of atmosphere as sunsets. Why aren't they as colorful?

Well, don't underestimate the time-of-day factor: for most humans, nearly everything looks better at 8:00 P.M. than at 5:30 A.M.

However, in this case you're right—on most days, sunsets are much more brilliantly colored than sunrises. It doesn't make sense, since most of the same conditions are in place for each—a sun in a low sky sending rays careening through Earth's atmosphere. So what gives?

Credit humanity. Or rather, blame us. During the day we generate the bulk of our traffic exhaust and factory smoke, which lingers long enough to give brilliance to our sunsets. During the night, much of the pollution gets blown away from population centers, so less of the blue light gets scattered, resulting in less intense reds and yellows in sunrises than in sunsets.

Send in the Clouds

Why are clouds white when water is clear and air is clear?

Clouds consist of millions of tiny water droplets. Light waves of all the colors bounce and reflect off and refract through the water droplets evenly, sending all of the various colors off in (more or less) equal measure. As you may remember from school, the "color" white is technically all of the colors mixed together. Because all of the colors are being evenly dispersed by the water droplets, clouds appear white.

How do clouds float in the air if they contain a lot of water?

Despite having a fluffy facade of unbearable lightness, clouds are actually pretty heavy things. Take even a small, fluffy cloud measuring a mere cubic kilometer. The water in that cloud would typically weigh about a million kilograms, about the same as 500 midsize automobiles. So how does all that weight stay up there?

Well, it turns out that the air in that same cubic-kilometer cloud is also heavy—about a thousand times heavier than the water is. The water vapor is less dense than the surrounding air, so the surrounding air exerts an upward force on the cloud, causing it to float and keeping it up in the air.

But wait, there's more. Clouds are usually formed in warm air that is moving upward, adding even more buoyancy. So, as long as the water is in the form of tiny droplets, it floats just fine in the warm up-currents. It's only when the air cools that the mist condenses into larger drops, and the downward pull of gravity trumps the upward push of rising air. Then rain falls, and the moisture that is the lifeblood of the cloud slowly gets sapped away.

How many miles up are the highest clouds?

Cirrus clouds, those high, wispy-looking clouds you see on a clear day, are above 18,000 feet—about 3.5 miles high. The highest clouds ever recorded were just under 60,000 feet, or about 11 miles high, though the instruments used to measure such things are less than precise. To put this in perspective, Mount Everest is 29,035 feet (5.5 miles) high, and the Boeing 747 is allowed by law to fly as high as 45,100 feet, or a little over 8.5 miles high.

What does cirrus mean anyway?

It means "curl of hair" in Latin. The four main types of clouds were all named in Latin, according to what they looked like to someone standing on the ground. Besides cirrus, *cumulus* means "heap," *stratus* means "layer," and *nimbus* means "rain."

All Hail Freezer!

How come it sometimes rains instead of snows, even though the temperature is below freezing?

There are several ways this can happen. It helps to think of the air—from clouds to ground—as a series of layers, each with its own temperature, and each with the ability to affect falling precipitation. For example, the cloud may be below freezing temperature, so it will drop snowflakes, but the snowflakes have to pass through other layers before reaching the ground, some of which might be above freezing. In that case, the snow can turn back into water and hit the ground as rain. Conversely, if the cloud is above freezing, it can drop rain that freezes on its way down through subfreezing layers of air, causing sleet. Finally, if the lowest level of air next to the ground is the only layer that's below 32 °F, the raindrops may not have time to freeze before they hit the ground. They will, however, freeze into ice upon landing, producing freezing rain. This is the stuff that can coat power lines and tree branches, sometimes bringing both crashing to the ground.

Is hail the same as freezing rain?

It's not even the same thing as sleet. Freezing rain–raindrops that freeze once they hit the ground–occurs only during wintry, subfreezing weather. Hail, on the other hand, almost exclusively falls during thunderstorm season between spring and early fall.

Here's how hail happens. When thunderstorm clouds gather together from heat and moisture, they often reach great heights, swirling up into low-temperature regions of the sky. The swirling warm and cold air rotates up and down within these cloud formations. Rather than falling normally, raindrops can get stuck in a cold, windy loop that occurs in the higher, colder regions of the storm. When this happens, they can collide with other freezing and thawing drops, forming an even larger drop. The longer a drop is in this loop cycle between the warm and cold areas of the storm clouds, the bigger it gets– attracting more water vapor, joining with other similar drops, freezing, partially thawing, and refreezing again. Finally the weight of the hailstone becomes great enough for it to fall out of the windy loop inside the storm cloud. It falls to the ground as a round and painful chunk of ice.

Raining Bats & Frogs

Where did the phrase "raining cats and dogs" come from?

Well, one rational theory is that it's just a comically overblown expression of hyperbole. However, other theories exist as well. Some believe it literally describes an actuality of the past.

Here's what we do know for sure. The phrase comes from England, and its first recorded use is by Jonathan Swift in *Polite Conversation,* written circa 1708 and published thirty years later. (An earlier variant, "rain dogs and polecats," appeared in 1652 in Richard Brome's *City Witt.)*

One possible explanation reads as such: In London, dwellings were packed closely together and the roofs were used by the stray animals of the city as highways and refuges. Some would die up there, and when a torrential downpour washed debris off the roofs, down would come animal carcasses, too.

Other word experts counter with their own pet explanations (as it were)—for example, that rain on metal and wood roofs sounded like a great brouhaha between the animals. Another theory is that in Norse mythology, cats influenced the weather, and wolves were the companions of Odin, the sky god. Still another cites the archaic French word *catdoupe,* meaning waterfall or cataract.

In other words, none of the experts really know for sure. And therefore, neither do we.

How is it that it once rained frogs in Mexico?

Not just in Mexico—it's been raining frogs all over the world, if you believe the reports. And not just frogs either, but fish, periwinkles, crabs, jellyfish, coins, worms, baby alligators, and even ears of corn. These are not just stories from the tabloids; some have even been reported in reputable scientific journals.

Throughout history there have been times that living (or formerly living) creatures have fallen from storm clouds. Probably the earliest recorded incident of animals raining from the sky occurred in ancient Greece some 1,800 years ago, when frogs rained down in such quantity that people fled the city.

It may sound miraculous, but there's at least one logical explanation. Storm winds are pretty spectacular things. They can form tornadoes and waterspouts that have enough force to suck even heavy objects up into them—frogs, fish, snails, puppy dog tails, and other small creatures are sometimes no match. The poor animals can then be dropped onto very surprised people below.

So yes, in 1997 a tornado made news in Villa Angel Flores, Mexico. It had apparently grabbed frogs from a pond and deposited them in a neighboring village, scaring the wits out of the town's residents.

How many thunderstorms can occur worldwide in a day? As many as 50,000.

What causes thunder and lightning?

Angry gods, of course. Haven't you ever watched cartoons?

Okay, it's electricity, and it's not that far removed from what happens when you rub a large, furry cat. Start with those tall, puffy cumulonimbus clouds that develop quickly as hot, moist air flies up into colder layers. The turbulence forces water vapor to move up and down at high speeds, rubbing a huge quantity of electrons loose from their molecular moorings and creating a massive buildup of electricity. Lightning is the result, and as the blinding bolt discharges, the 40,000-degree heat expands the surrounding air, creating a big bang of sound waves that we hear as thunder.

How can you tell how far away a lightning strike is?

The speed of light is a lot faster than the speed of sound—about a million times faster—so lightning can often be seen before thunder is heard. Next time you're in a storm, wait until lightning flashes, then begin timing the seconds until you hear thunder. Divide the number by five, and that's about how many miles the storm is from you. Be aware, though, that if you can hear the thunder, the storm is close enough to be dangerous, so don't hang around looking skyward with a stopwatch in hand. If the thunder crashes as lightning flashes, take heed—the storm is directly on top of you (see below).

What's the difference between thunderstorm warnings and thunderstorm watches?

The same as a tornado watch and warning: A thunderstorm watch means that all the ingredients are in place to possibly create a severe storm. A warning means a storm of this magnitude has already been spotted, so take cover now!

What should you do when you're caught in a lightning storm?

Lightning is an electrical charge trying to find its way into the ground. It will use whatever conduit it can to best accomplish that. Air doesn't conduct electricity all that well. Actually, it's pretty resistant, so when there are other alternatives, lightning will use them to make its way toward ground.

The best thing to do is to get inside a building. However, that's not always a possibility.

Because metal is a great conductor of electricity, lightning will travel through it whenever it gets the chance, so stay away from metal poles, set down your golf clubs, and take off your jewelry. Wood's not a bad conductor either. Therefore, don't stand under a tree, especially one that is off by itself in an otherwise clear area. Lightning will try to find the easiest conduit, which of course is usually the tallest object in the area. If there's no place to get inside, get down. Try to find the lowest, driest point in the area, and get as low as you can, duck-and-cover style, with your hands and arms over your neck and head. Finally, if you're with others, get away from them. Your group members should spread out to at least twenty feet apart. The more of you there are clumped together, the better conduit your group will be for a lightning bolt.

Can the rubber tires of my bicycle offer the same protection from lightning as a car's rubber tires?

No. It's true that a car is a relative safe haven during a lightning storm, but it's not because of its rubber tires. As a matter of fact, rubber offers no significant insulation whatsoever from lightning. Cars work as an okay shelter from lightning storms because the metal frame will tend to attract lightning and conduct its electricity through the metal and around the passenger compartment, as long as you don't touch metal and keep your windows closed.

The same concept applies to a bicycle. Well, everything except keeping the rider safe. If a lightning bolt hits your bicycle frame with you sitting on it, you will find yourself very painfully part of the circuit.

If you're on a bicycle when a lightning storm hits, the best thing to do is get off, park it some distance from where you're going to be crouching duck-and-cover style, and hope that the lightning chooses it instead of you, if such a choice is to be made.

What are my odds of getting struck by lightning?

Generally, your odds are 3,000,000 to 1 that you won't get hit. But in lightning strikes, as in real estate, location is everything.

If you're a park ranger and spend most of your time outdoors, your odds of course will be higher than if you're a practicing couch potato. If you get your exercise as a golfer, your odds of being hit are substantially greater than if you're into jazzercise. Pro golfer Lee Trevino, for instance, has been struck. As a result, when rain rolls in, Lee runs. What was it like? According to Trevino: "There was a thunderous crack like cannon fire and suddenly I was lifted a foot and a half off the ground. 'Damn,' I thought to myself, 'this is a hell of a penalty for slow play.' "

> What percentage of lightning fatalities happen on a golf course?
> Twelve percent.

The Full Spectrum

Are all rainbows the same shape?

Yes. All rainbows are completely round. It's just that we don't usually see the whole rainbow because we're standing on the ground and looking toward the horizon. Unless you're in the air, or looking down from a mountain, you can usually just see the top part of a rainbow. The "bow" in rainbow refers to this arch.

When's the best time to see a rainbow?

If the sun is too high in the sky, you won't see one. The *only* time you can see a rainbow is either in the early morning or in the late afternoon. There's a reason for this. For you to see a rainbow, the sun has to be in back of you and shining through the rain. If the sun's too high, any rainbow that might form would end up projected below the horizon, making it invisible.

What's the order of the colors in a rainbow?

From top to bottom: red, orange, yellow, green, blue, and violet.

Except in the case of double rainbows, which occur when light reflects off another set of raindrops and forms an outer rainbow that's a pale imitation of the original. The color sequence in the reflected rainbow is reversed—blues on the outside, then yellows, with the reds forming the inside of the arch.

No Business Like Snow Business

How much water does it take to make an inch of snow?

If it's average snow—not too wet, not too dry—you'd need about a tenth of an inch of water to make an inch of snow. The general rule is that one inch of water produces ten inches of average snow. But that's the average. An inch of water would make about fifteen inches of dry and fluffy snow, but only five or six inches of wet and soggy snow.

How can you survive an avalanche?

Hope for a small avalanche instead of a big one and wish for luck. A bona fide large avalanche can exert so much pressure that it snaps trees, rocks, and bones in its path, carrying enough snow to bury twenty football fields in ten feet of it. Large or small, you'd think an avalanche would kill about everybody it strikes, so the encouraging news is that one in three people caught in an avalanche actually survives.

If you happen to find yourself in the path of falling snow, there are a few things you can do to improve your chances of survival. The right equipment is your first line of defense. A small shovel and a long probe can mean the difference between suffocating to death under several feet of packed snow and being seen and rescued. In mountainous areas, always carry a transceiver, turning it to transmitting mode so a partner can possibly locate your signal if you end up caught.

The National Snow and Ice Data Center Web site at http://nsidc.org carries step-by-step instructions on how to react:

> Yell and let go of ski poles and get out of your pack to make yourself lighter. Use "swimming" motions, thrusting upward to try to stay near the surface of the snow. When avalanches come to a stop and debris begins to pile up, the snow can set as hard as cement. Unless you are on the surface and your hands are free, it is almost impossible to dig yourself out. If you are fortunate enough to end up near the surface (or at least know which direction it is), try to stick out an arm or a leg so that rescuers can find you quickly.

If you are in over your head (not near the surface), try to maintain an air pocket in front of your face using your hands and arms, punching into the snow. When an avalanche finally stops, you will have from one to three seconds before the snow sets. Many avalanche deaths are caused by suffocation, so creating an air space is one of the most critical things you can do. Also, take a deep breath to expand your chest and hold it; otherwise, you may not be able to breathe after the snow sets. To preserve air space, yell or make noise only when rescuers are near you. Snow is such a good insulator they probably will not hear you until they are practically on top of you.

Above all, do not panic. Keeping your breathing steady will help preserve your air space and extend your survival chances. If you remain calm, your body will be better able to conserve energy.

Extremes & Nightmares

Where's the hottest place on earth?

Officially, it's Death Valley, California. On July 10, 1913, temperatures reached 134 °F (56.7 °C). Four years later, during the summer of 1917, Death Valley broke another record–the highest average temperature–by maintaining temperatures of over 120 °F (48 °C) for forty-three consecutive days.

These are the official world records because they have been thoroughly documented. However, we can assume that other places–particularly in Africa–have likely reached higher numbers. For example, on September 13, 1922, Al Aziziyah, Libya, reached a reported temperature of 136.4 °F (58 °C), according to the National Geographic Society, but it was never officially recognized by the world's weather services, so the better-documented Death Valley record still stands.

Having said that, Dallol, Ethiopia also deserves mention. It holds the record for the highest average annual mean temperature in the world. From October 1960 to December 1966, the average mean temperature was 94 °F (35 °C). Yep, that's some mean temperature.

What's the coldest climate where humans live?

Oymyakon, Russia, has a substantial, ongoing population of 4,000, and its temperatures have fallen as low as −72 °F (−58 °C). People have visited spots even colder, though. In Vostok, Antarctica, people were present when temperatures got down to −89 °F (−67 °C). But no one has set up residence there yet. Too bad—we suspect the real estate prices would be quite reasonable.

Is Australia the driest continent on Earth?

Surprisingly, no. While much of Australia is very dry, the driest continent is a desert you might not even think of as a desert.

It's Antarctica. Deserts, you may remember, are not necessarily a place of great heat, but are instead characterized by a lack of rainfall. Although Antarctica has a lot of frozen moisture on the ground, it's not because the continent gets a lot of snow each year. It's because the little snow it gets rarely melts. On the average, the frozen continent gets less precipitation in a year than most places in North America—about four inches of snow per year.

So Antarctica's the driest continent overall. However, the single driest location on Earth is not in any of the famous deserts you'd think of—not the Gobi Desert, the Sahara, or the Mojave Desert. It's the Atacama Desert in Chile, which gets less than .1 millimeters (.004 inches) of rain per year, on average. Some years it sees no rainfall at all.

In contrast, Mawsynram, Assam, in India is perhaps the wettest place on Earth, seeing an average of 11,873 millimeters—or 467.4 inches—of rain every year, more than an inch of rain on an average day. (In comparison, Cleveland, Ohio, gets 36.6 inches of rain in an average year.)

Does the equator have any seasons besides summer?

Although locations along the equator are hot throughout the year, they do still have seasons. Usually two: a wet season and a dry season.

Discomfort & Joy

What's windchill? Does it lower the temperature?

The windchill index is an interesting thing, an attempt to take a scientific certainty—temperature and wind speed—and quantify its effect on the perceptions of humans. In other words, it calculates perceived temperature. Scientists working in Antarctica developed the first formula to determine when the wind will turn a frigid temperature into a deadly one. The official windchill tables were altered again in 2001.

In essence, the windchill index measures the cooling effect of air blowing past you along with the actual temperature. For example, if you walk outside in a windless 30 °F, there are no surprises. It feels like 30 degrees, and your body cools and reheats itself at a constant rate. However, if the wind is blowing at 10 miles per hour, that 30 degrees will feel like 16 degrees. In 30 mph winds, that 30 degrees will feel like −2 degrees. As wind increases, it blows your body heat away, making the temperature seem even colder.

The windchill index was such a huge hit with the public that scientists decided to find a similar measure that would circumvent all of those banal summertime "It ain't the heat, it's the humidity" conversations. In hot weather, the heat index is a similar attempt to quantify the subjective perception of heat when mixed with varying levels of humidity, based on the speed that sweat evaporates from the body.

My car's antifreeze will go down to 0 degrees, but what if it's 10 degrees with a windchill factor of −15?

Not to worry. Windchill applies only to heat-generating people and animals. No matter how hard the wind blows, your car's temperature will not get below whatever the actual temperature is.

Where can I see a table of the windchill and heat indexes?

 Ask

Where does the phrase "the dog days of summer" come from? What does it mean?

The traditional "dog days" of summer fall between July 3 and August 11, noted for their extreme heat and humidity. During this time of year, the star Sirius is at its brightest and can be seen rising alongside the sun. The phrase actually dates back to the Egyptians. They believed that the star gave off extra heat and humidity to augment the already formidable heat of the sun. Sirius is the "dog star" from the constellation Canis Major (Latin for "Big Dog"), hence the name.

Wasn't "Indian summer" meant to be a derogatory term?

Maybe, or maybe not. The phrase, of course, refers to a time in late fall when the weather turns unseasonably warm and mild. For many years, it was believed that the name came from the idea that Native Americans can't be trusted, that "Indian summer" was a time when settlers were lulled into a false sense of security before winter snuck up and massacred them in their sleep.

Some revisionist weather historians argue, however, that the term came from a more benign association between Native Americans and the smoky last days of autumn. The earliest written reference to the term was found in a journal entry written in the 1770s by a French immigrant farmer in the American colonies. He wrote that an "Indian summer" was "a short interval of smoke and mildness." He may have been referring to two last-minute prewinter preparations by the Native Americans that generated a lot of smoke. For one, Indians would often use fires in late autumn to smoke prey out of its hiding spots, in order to make sure they had enough meat stockpiled for the long winter months. Secondly, some Native Americans also burned portions of their lands to kill weed seeds, remove the leftover crop stubble, and return nutrients to the soil for the next year's crops.

Then again, maybe it's just conversationally safer to avoid the references to "Indian summer" and stick with "How about this weather we're having?" generalities.

If you're standing on top of a mountain, and are that much closer to the sun, how come you're colder than when you're at sea level?

Although it sounds like a lot, and certainly seems like a lot if you're the one doing the climbing, even Mount Everest is only 5.5 miles high. Considering that the sun is 93 million miles away from Earth, a measly 5.5 isn't going to largely effect the amount and intensity of the sunlight you're receiving. To see any significant rise in temperature due to distance to the sun, you'd have to get into a spaceship and set the controls for the heart of the sun—perhaps stopping after 10,000 miles or so.

That answers why you aren't any warmer; here's why you're actually colder. Although a few miles up makes little difference in terms of sun radiation, it makes a whole lot of difference in terms of atmosphere. For one thing, much of the sun's heat comes from striking Earth and being radiated and reflected back up. On a mountain peak, you don't get much of either of those things. Also, as air moves upward, it expends more energy and becomes cooler in the process. Additionally, you have to consider that the higher you climb, the thinner the air, and thinner air holds less heat. And let's not forget the fact that the insulating, low-lying clouds that trap heat close to the ground are now going to be below you instead of above you. Finally, the angle of the sun also plays a role in cooling mountains. Sunshine coming at an angle spreads its energy over a greater surface area. A mountainside has a slanted surface area, spreading its solar energy over a larger area. Add the additional cooling properties of the residual snow and ice at the mountaintop, and you have a pretty chilly result.

If Earth's getting hotter, Jeeves, then how come the winter temperatures are lower?

Actually, they're not. Over the last many years, the world has slowly gotten hotter. This difference in temperature is minuscule—a fraction of a degree—which means that to most of us, things seem normal. However, it's a potentially ominous indicator of global warming. It doesn't take much of an increase in average temperatures to have a devastating effect on humans, crops, plants, and wildlife worldwide.

Not that this warming trend precludes individual locales or regions having unusually cold winters. You may have heard your local weather guy talk about record-breaking low temperatures, or the worst snowstorm on record in your city, but these isolated events are exceptions to the general trend. The ten hottest years on record have all occurred within the last two decades.

Is That Celsius or Fur-enheit?

How accurate is the official groundhog, Punxsutawney Phil?

Over the last sixty years, the average accuracy rate for groundhogs on Groundhog Day has only been about 28 percent. Punxsutawney Phil's got a slightly more impressive record—he predicts correctly about 39 percent of the time.

Maybe they're using the wrong animal for the job. The original "groundhog" wasn't a groundhog at all, but a badger. The tradition is a German one, and when German immigrants came to America, they couldn't find a stinkin' badger in all of Pennsylvania (at least not any that looked like their Old World variety), so they had to settle for the much more common groundhog.

How can you use a cat to tell what the temperature is?

How accurate a read are you looking for? A cat can be used to give a very (very) general idea of temperature. The colder the temperature, the more a cat curls around itself to sleep. If your cat's lollin' around, stretched out on her side or back, you can bet it's warm. If said cat is tightly curled, with face buried and tail wrapped as tightly around her body as possible, she's cold. In other words, get yourself a sweatshirt.

How can you use a cricket as a thermometer?

Well, first of all, don't stick it under your tongue. That won't give you the information you seek, and nearly always damages the insect.

First, some background: Cricket metabolism slows at a consistent rate when they get cold and speeds up when they get hot. This is true of not just crickets but all cold-blooded creatures. However, crickets make especially good thermometers because they chirp regularly and incessantly. The intervals between chirps become an indicator of temperature—on cold nights, they slow down; on hot nights, they speed up. If you count the number of cricket chirps and do some simple math, you can get a moderately accurate sense of the temperature. Here's the simplified formula first, the one taught in nature class: To get a close approximation of Fahrenheit temperature, count the number of chirps in fifteen seconds and add 40. If you prefer your crickets on a metric scale, count the number of chirps in a minute, add 50 and divide by 9.

But what if an estimate isn't good enough for you, and you want a more precise measurement? Well, then, you have to calibrate your instrument. First of all, determine what kind of cricket you're using. All of the results below give you the temperature in Fahrenheit:

- If your cricket is black, it's a common field cricket. Count its chirps for fifteen seconds and add 38.
- If your cricket is small and pale green, and you found it on a tree, it's a tree cricket. Count the number of chirps for seven seconds and add 46.
- If it's white, and it's in a tree, you're in luck—you've got a snowy tree cricket, the most accurate cricket of all. Count its chirps for fourteen seconds and add 42.

Just one final warning—the cricket thermometer is only accurate in summer weather, not in the chill of other seasons. For example, if there's snow on the ground and you hear zero chirps out there, you can't do the math and conclude that the temperature is 40 degrees.

Blowing Hot & Cold

How come hurricanes hit the East Coast but not the West Coast?

There are two really good reasons.

The first is temperature. A hurricane is created and fed with the hot air that rises off warm water. Conversely, cold water's cooler air tends to sap the energy from a hurricane, pretty quickly killing it dead (the waters along our eastern coast have just returned from a tropical vacation before being routed north, so they're comparatively warm even as they flow past New York City). The waters off the West Coast, on the other hand, have flowed south after being chilled in the Arctic, and so their average temperatures are 20 degrees colder than water along the East Coast.

The second reason is that hurricanes in the Northern Hemisphere tend to travel in a west-northwestern pattern. Look at a map, and you'll see that southeast of America's East Coast is a huge expanse of tropic waters. Southeast of America's West Coast are states like Arizona and New Mexico. Since hurricanes can't start over dry land, Californians can for the most part refrain from worrying about big winds and invest that nervous energy into concern over earthquakes and such.

True, sometimes a big storm produced in western waters will occasionally surprise everyone by making a U-turn and heading back toward the beaches of Santa Cruz and Malibu. But because their power gets quickly sapped by the cool water, they almost never get close enough to scare the surfers.

What's the difference between a typhoon and a hurricane?

There isn't much difference, beyond their names and locations—they're both tropical cyclones. Typhoons are hurricanes that occur in the western part of the Pacific Ocean or over the Indian Ocean. Hurricanes always take place on the Atlantic Ocean or the Caribbean Sea.

Hurricanes get their name from the Spanish *huracán,* taken from the Taino Indians' *hurákan,* which they might've gotten from the Arawak Indians, who have a word *kulakani* that means "thunder." Typhoons get their name from *tai-fung,* a word in a Chinese dialect that means "great wind."

I've heard that wind shear can knock an airplane out of the sky. What is wind shear?

Wind shear is a fancy name for turbulence—that windy stuff that makes airborne airplanes bounce a little. Wind can sometimes change directions suddenly, either vertically or horizontally, creating eddies and swirls and bumping into other wind currents headed in the opposite direction. Needless to say, if you're floating on a cushion of air, this sort of disruption is scary and sometimes dangerous.

The wind shear responsible for airplane crashes generally comes from wind gusts pushed downward and outward by severe storms. These winds are much faster, harder, and less predictable than your usual turbulence, and its not a bad time to avoid going into the air and being tossed around by their strength.

In the movie Twister, *they sent a sensor up into a tornado. Do meteorologists use these often? How do they work?*

Offscreen, sensors like that don't exist, although the idea has been experimented with by the National Severe Storms Laboratory, a government group that tracks and researches (you guessed it) severe storms. Here's what the NSSL's Web site FAQ says regarding the movie devices:

The movie TWISTER was based upon work NSSL did in the mid-1980's using a 55-gallon drum filled with various meteorological sensors. It was called TOTO (TOtable Tornado Observatory). NSSL tried for several years to put it in the path of an oncoming tornado, but had minimal success. It did not have the sensors that fly up into the tornado, like in the movie. However, that is not a bad idea and with the advances being made in computer technology, we might be able to do that someday.

What does a tornado sound like?

The sound, say those who've experienced it firsthand, can be compared to a hundred freight trains traveling nearby at the same time.

I remember hearing about a way to use your television as a tornado warning device. Do you know how to do it?

We've never actually tried this, so we can't completely vouch for it. And of course, it's best to make sure you're in a safe place before trying it—we'd hate for you to be standing in front of your home entertainment center upstairs instead of huddled in your basement during a tornado warning.

Having said that, here are the instructions as we understand them. First tune your television to Channel 13 (over the air, not cable) and darken the screen to almost black using your television's brightness control. You're using Channel 13 as a control screen, because it's said to be the VHF channel least affected by storms.

Next, turn to Channel 2. On the other end of the VHF spectrum, it supposedly is the most susceptible to storm interference. Lightning will produce momentary colored bands of varying widths across your darkened screen. Ignore them. Watch instead for the screen to suddenly go totally bright. According to those who claim this system works, this massive interference on your screen means that there's a tornado within fifteen to twenty miles.

Toying with Science

It's time to stop thinking of toys as mere playthings. From the statistics and icosahedrons of the Magic 8-Ball to the gyroscopic precession of the boomerang, sometimes toys are the best things around to demonstrate the principles of science.

Reply Hazy . . .

What're the odds of getting a positive answer from a Magic 8-Ball?

"MOST LIKELY." Out of the twenty possible answers inside a Magic 8-Ball, ten of them are positive, five are negative, and five are noncommittal, suggesting you ask again later. That means you have a 1 in 2 chance of getting a positive answer, and only a 1 in 4 chance of getting a negative one, with another 1 in 4 shot at a neutral answer.

The possible affirmative answers are SIGNS POINT TO YES, OUTLOOK GOOD, YOU MAY RELY ON IT, MOST LIKELY, YES, YES DEFINITELY, IT IS CERTAIN, WITHOUT A DOUBT, and AS I SEE IT, YES. The negatives are MY REPLY IS NO, DON'T COUNT ON IT, OUTLOOK NOT SO GOOD, VERY DOUBTFUL, and MY SOURCES SAY NO. The noncommittals are CANNOT PREDICT NOW, ASK AGAIN LATER, BETTER NOT TELL YOU NOW, CONCENTRATE AND ASK AGAIN, and REPLY HAZY, TRY AGAIN.

What shape is the answer device inside a Magic 8-Ball?

Although Mattel, the toy company that now owns and distributes the toy, calls this patented contraption the "20-sided answer cube," there is a more mathematically accurate name for the piece. All twenty-sided polyhedrons are called icosahedrons.

Bossy Died for Your Burnt Sienna

If they're just made of wax, why do crayons smell?

Have you ever wondered what stores do with beef that's past its freshness date? It's sold off to various manufacturers who render the fat (called tallow) and put it into all kinds of products, from hair conditioner to—you guessed it—crayons. The fat is often called "free fatty acid" or "stearic acid" on labels. That obviously sounds more appealing to manufacturers (and consumers) than "beef fat."

Other products made with beef byproducts include asphalt, explosives, lipstick, photographic film, shampoo, detergents, fabric softener, and candles, to name just a few.

If you have an ethical problem with meat byproducts being in your child's drawing utensils, you can purchase crayons made of beeswax or vegetable products, such as soybean oil.

Where can I find vegetable crayons?

Just by regular use, is there one crayon that usually gets up first from a box of crayons?

Definitely. Almost without fail, black's the first to go. This makes sense; folks of all ages tend to use black as an outline and background color for furniture, lettering, buildings, etc. Less predictably, red is usually the second one to go.

In the Groove

What were phonograph records made from?

The master cut of a record was usually made from a layer of lacquer on top of a flat aluminum plate. A machine with a piece that looks a lot like a record-player stylus and needle was fed sound vibrations. In turn, this apparatus cut grooves with dips and bumps that correlated to the vibrations into the surface of the master cut, making that recognizable spiral on the surface of a record most of us are familiar with. The master cut would then be used to stamp out numerous copies that would be sold to the public. The copies (records) themselves, though, are a mixture of polyvinyl chloride (PVC) and, strangely enough, free fatty acids from beef tallow (see above).

How does a See 'n Say toy work?

In a world where computers and chips can create everything from dancing monsters to singing fish, it's nice to know there are still some tried-and-true toys that are based on technology created over a century ago. The See 'n Say basically uses the same technology that a record player does. There's a disk inside the outer plastic shell that has sounds recorded on it with grooves cut into the surface like a record. A needle "reads" the grooves on the disk, and transfers the sound into a speaker.

When you choose the picture on the outside, then pull the string, the needle is pulled off the surface of the disk as the disk gets wound up. Letting go of the string spins the disk, positions the needle above the correct place on the disk, and drops it so sound will play.

Perchance to Dream

How does a yo-yo "sleep" at the bottom of the string without coming back up?

First, a quick rundown of the anatomy of a yo-yo: two disks of equal size are placed together and connected with a little part in the middle called an axle. A string is affixed to the yo-yo, either by tying a loop around the axle or by attaching the end of the string directly to a point in the axle. A yo-yo is able to "sleep," or stay spinning at the bottom of the string, because as the string loops around the axle, gravity keeps it down instead of winding the yo-yo back up.

The axle has to be smooth enough that it can spin freely in the loop of the string, yet has to have just enough friction that when you jerk the string, the axle will grip it and wind back up. When you want the yo-yo to come back up, a simple jerk will begin the process of the string wrapping around the axle. Of course, none of this works at all–the yo-yo will not "sleep"–if the string is affixed to only one place on the axle.

Would a yo-yo "sleep" in zero gravity?

On April 12, 1985, NASA decided to test just that. It sent a yo-yo into space with the astronauts aboard the space shuttle *Discovery.* What they learned was that a yo-yo would indeed work with little-to-no gravity. With a gentle push, it glided down the string and back up again. Simply dropping the yo-yo, though, didn't work; it had to be thrown. Furthermore, they learned that without gravity, a yo-yo could not in fact "sleep."

Flipping a Disc

Does the spin of a Frisbee make it hover?

Because the tops of flying discs are rounded, air goes over them faster and has less density, creating lift, as with airplane wings. Air gets trapped underneath the disc and pushes upward, slowing the fall rate. The thrust of the throw gets the lift process started, but the spin only adds stability and keeps the disc from wobbling or tipping.

Right Back Atcha, Babe!

How does a boomerang come back to you?

As one boomerang physics Web site states it: "Magic makes it come back." Or scientifically, it's a combination of lift, spin, and something confusingly titled gyroscopic precession. Gyroscopic precession is the principle that states that if an object is spinning, and movement (the tilt) is exerted on the object, it will move at right angles to the direction it is being tipped.

The blades of a boomerang are slanted, like helicopter or airplane blades. The wind travels across the top and creates lift, allowing the boomerang to fly. Meanwhile, the first blade that heads into the wind creates more lift than the other blade does, causing a tilt to the boomerang. The tilt that the lead blade produces on the boomerang causes the curving arc that brings the boomerang back to the thrower. The strength of the throw as well as the lift that the blades create generate the momentum the toy needs to make the entire flight. At least, if you throw it right in the first place!

Where else can I see gyroscopic precession in action?

Ask

The Red, Red Robin Keeps Bob, Bob, Bobbin'

What makes the Dippy Bird tilt down to drink and then pop up again?

Have you ever noticed how fast paint thinner evaporates at room temperature? It's this little fact that makes the glass and plastic dunking birds work.

Here's how it works: The bulb that holds the colored liquid (methylene chloride) inside has a bubble of vapor on each end. At room temperature the liquid congregates mostly toward the bottom, with the vapor evenly balanced on both ends. However, when you wet the head, the clothlike flocking allows the water to evaporate. The evaporation cools the head, reducing the pressure of the vapor on that end, which then sucks the liquid up the bird's neck. Eventually the bird becomes top-heavy, so the head plunges back down into the water glass. This water is warmer than the evaporation-cooled head, which expands the vapor there, pushing the liquid back toward the bottom of the body. But this doesn't last long, as evaporation on the head gets going yet again, making the bird "drink" without stopping.

Up, Up & Away!

Why does a helium balloon float, but a regular one falls down?

Anything that's lighter than air will float. A liter of air at sea level weighs about 1.25 grams; a liter of helium, though, is only .18 grams. Therefore the air sinks below the helium balloon, causing it to fly upward.

Higher & Higher

Why does breathing helium make you sound like a munchkin?

This seems like it should be an easy one to answer, but you'd be surprised at the explanations we've heard from people who should know better. Let's go in order of what we consider least reasonable and work our way up:

1. **Because helium is less dense, sound travels faster through it, so the wavelengths are shortened. Your ears hear the voice as being higher.** You can find this explanation on many science Web sites and in many books. This seems plausible enough on first read—after all, air is almost seven times denser than helium. However, the idea falls apart with a little more thought. For one thing, your voice would only travel through a few inches of helium from the larynx to the mouth before hitting regular air—not enough to raise the pitch significantly. Furthermore, this

theory takes a larger hit when you consider the experience of crew members on deep-sea submarines. To avoid the problem of the "bends"—nitrogen bubbles in the bloodstream—deep-sea crew members breathe an atmosphere of helium mixed with oxygen. It's true that their voices sound higher, but that should also mean a higher pitch for all other sounds as well. In truth, most sounds—footsteps, the clatter of dishes, taped music, etc.— are unchanged in a helium atmosphere. So let's just say this theory doesn't hold water.

2. **When sounds coming from a helium-filled throat hit the dense air outside, the wavelengths compress like traffic coming to a slow spot in the road. Your ear hears these shortened wavelengths as a higher pitch.** This theory is elegant and plausible. The problem is that it's easily sunk by the same deep-sea submarine as the last theory. If it were right, crew members in a helium-based atmosphere shouldn't have high voices—but they do. Also, this would strongly imply that breathing normal air from a balloon and talking within a helium atmosphere would make your voice sound unnaturally low by reversing the process. Tests have shown that this is not the case; people breathing in regular air in a helium atmosphere have normal-sounding voices.

3. **The reduced density of helium has less wind resistance, allowing the vocal cords to vibrate faster than they would in air. This makes your voice higher.** In the same way that less wind resistance would allow a fan to rotate faster, for example, or a tennis racket to swing faster, it sounds plausible that helium's lower density would allow the larynx to vibrate faster and so sound higher, regardless of what the surrounding atmosphere contained. This seems to be consistent with the submarine test above, so we'll stick with it until we hear something even more definitive.

One last fun fact: Did you know that breathing helium before playing a clarinet also makes the clarinet's pitch higher? Research on other instruments is still pending. Maybe soon we'll see a high-pitched helium-powered band.

Enlightenment with a Snap

How do glow sticks work?

The chemical process used in a glow stick is not that far from the chemical process that occurs inside a lightning bug. When you open the glow stick package, you find a stick with yellow liquid inside (generally, a chemical called luciferin). Inside the yellow liquid is a glass tube that contains a chemical enzyme (generally, luciferase). When you break the glass tube by bending the stick, per the directions, the two chemicals mix, creating a lot of energy and giving you a good glow. A lightning bug creates a glow in the same way—mixing the two chemical enzymes luciferin and luciferase.

Mood Indigo

What makes a mood ring change colors?

Thermotropic liquid crystals. They're little crystals that change molecular states when the temperature changes. When their molecular states change, the crystals' ability to absorb light also changes, meaning the color you see will change, too.

The ring picks up body heat from your finger, and transfers it to the crystals in the ring. The crystals are calibrated to reflect the color green at a surface temperature of about 82 °F. When the surface temperature gets cooler, they go darker and then gray/black. When they heat up, the color changes to blue hues.

Does the ring actually show your emotions? Well, not exactly. It only reflects the small differences in surface temperature on your hand. However, the circulation in your hands will generally tend to increase when you're relaxed, increasing the surface skin temperature, and will sometimes decrease with tension, lowering the surface skin temperature. Based on that change, the mood ring instructions tell you to read your ring like this: Dark blue—romantic, passionate, happy; blue—completely soothed; blue/green—relatively calm;

green—normal; brown/
amber—slightly uptight; gray—
uptight; black—completely
wound-up and stressed.

Help, Jeeves! Where
can I find an
interactive, on-line
mood ring!

More Bounce per Ounce

How does a Superball bounce so high?

Wham-O, the company that brought the world the Superball,
starts with a good, bouncy rubber—in this case, Zectron, which
is Superball's trademarked name for the synthetic rubber
polybutadiene with some sulfur added for reinforcement. They
take that and add a whole mess of pressure when they mold the
ball. How much pressure? The company says it's about a
thousand pounds per square inch as they heat the rubber to
more than 320 °F.

What is Silly Putty made from?

Silly Putty, "the toy with one moving part," is a mix of boric
acid and silicone oil.

Squaresville

How many ways are there to solve a Rubik's Cube?

Well, in a sense there's only one, but according to the Rubik's
Cube official Web site, there are about 4.3 times 10 to the 19th
power, or 43,252,003,274,489,856,000, possible combinations.
With so many possibilities, you'd think it would be a lot easier!

Joker's Wild

How many possible hands are in a game of five-card poker?

Mathematically, there are 2,598,960 five-card hands possible
with a standard fifty-two-card deck. But that's child's play,
considering there are 635,013,559,600 possible hands in a game

of bridge. Still, these can't hold a candle to chess: a mathematician once calculated that the first ten moves in a game could be played in 170,000,000,000,000,000,000,000,000,000 different ways.

Tiny Bubbles

How do they get smells into Scratch-n-Sniff books?

Here's how it works: the oily extracts of the smells are placed inside really tiny bubbles of plastic—millions of microscopic bubbles. How do they do this? By emulsion—the mixing of oil and water. When they furiously mix the extracts with water, the oil is broken up into very tiny droplets, at which point a gelatin mixture is dropped into the oil and water. The gelatin settles around the oil droplets and forms a casing. The water is rinsed away, leaving gelatin-enclosed bubbles of smells. These bubbles are then mixed with an adhesive, and the adhesive/bubble slurry is printed onto paper.

When the book makes its way into your hands and you scratch the surface, you burst some of the bubbles and release the odor. Because there are so many on the surface, it's hard to actually scratch them all away, which is why you can pick up a twenty-year-old Scratch-n-Sniff book at a secondhand store and still use it.

The process of producing the bubble-coated fragrance is called microencapsulation, and it was first used for carbonless copying paper, but instead of fragrant extracts, ink chemicals were placed into the coated bubbles. When you place pressure on the paper with a pen point, the bubbles burst. The ink in the bubbles reacts with the chemicals in the bottom sheet of paper, producing a dark mark.

What's a good recipe to make bubble liquid for blowing bubbles?

Most of us have an inexplicable fascination with bubbles. Even Mark Twain once noted, "A soap bubble is the most beautiful thing, and the most exquisite in nature. . . . I wonder how much it would take to buy a soap bubble, if there was only one in the world." Fortunately for us, there's not. Here's a simple recipe you can use to make your own great bubble juice:

- Two cups of water
- One ounce of either Joy or Dawn liquid dishwashing detergent (these brands make the best bubbles)
- One ounce of plain glycerin (makes bubbles last longer by thickening the skin of the bubble. But don't use rose-scented–the alcohol-based scent weakens the strength of the glycerin)

For variations in size, try blowing bubbles through the back of a lawn chair, a straw, or a coat hanger–anything with a hole will work. Good luck!

Heavy Metal Fun

Is Slinky an ancient toy?

It would be cool if the story of the Slinky went back to the ancient Phoenicians, or Thorg, a druid from the Iron Age who discovered the property of a metal coil on stairs, but alas, Slinky's history is shorter than that. It was during World War II that a marine engineer for the U.S. Navy–Richard James–invented the first Slinky while working in a shipyard.

James was working on inventing quick-response antivibration devices for maritime instruments, trying to counteract the destabilizing force of ocean waves. One day when one of the springs got knocked off the table he was working on, it walked end-over-end down shelves and piles of books and onto the floor. In 1940, after the war was over, James's wife, Betty, coined the name "Slinky," they marketed his springs as toys, and the rest is history.

How does a Slinky walk down stairs?

First of all, a lot is owed to the stairs (or stack of books, or shelves) themselves. If the stairs you used were taller, the Slinky would flip over and end up rolling the rest of the way down. If the stair were shorter, the initial fall off the first step wouldn't give the toy enough energy to continue its cool walk down.

That said, the simple answer is that when a Slinky is poised on the top step and you gently nudge the top portion over the brink, you place energy into the coil that transfers, coil by coil, down the length of the wire. That energy moves in waves, like vibrating ripples, as the toy is pulled by gravity. The momentum of the toy's movement bounces it over and down onto the next step, and the "walking" continues.

Powder to the Peephole

What's the silver stuff in an Etch A Sketch?

It's the metal aluminum, ground so finely that it sticks to everything. This is true even if it manages to escape the insides of the Etch A Sketch. This powder inside the toy is peppered with little plastic beads that keep it from clumping. The window of an Etch A Sketch is made of glass, but it's painted with a plastic coating to strengthen it, and to keep it from shattering if it breaks.

Physics, Chemistry, & Math 101

Where No Man Has Gone Before

I heard a discussion about space travel in which somebody was talking about "Roddenberries." My dictionary doesn't have the term listed. What's it mean?

The *Star Trek* television show was created by a man named Gene Roddenberry. In tribute to the man and his promotion of far-ranging space travel, some *Trek*-loving scientists have coined the term "Roddenberry" to mean the distance traveled at the speed of light during a "traveler year."

A traveler year is an Earth year as perceived by a traveler going the speed of light. Since the theory of relativity says that time slows down substantially when you're traveling at that speed, a "traveler year" is the equivalent of 70.7 percent of a normal Earth year. That means that when Earthbound folk would be ringing in the New Year, a space traveler going the speed of light would be preparing to ring in September 16.

A light-year equals the distance that light travels in an Earth year–about 5.88 trillion miles. Thus, a "Roddenberry" is equal to .707 of a light-year, or 4.157 trillion miles.

Pennies from Heaven

If you dropped a penny off the top of the Empire State Building, would it leave a hole in the sidewalk?

We've all heard the story of the penny falling from the top of the Empire State Building and slicing its way through a parked car (or a pedestrian) below. It never happened, no matter what Jimmy Doofus in the fifth grade may have told you. Have you ever dropped a coin into a wishing well or fountain? You'll notice that it doesn't drop straight down to the bottom, but zigs, zags, and flutters all over the place, often landing quite a distance from where you thought you were dropping it. The same thing happens in air. Because of air resistance and the penny's not-very-aerodynamic shape, a dropped coin flips, dips, and spins during a free fall, slowing the speed of its fall substantially. In fact, pennies fall slowly enough–just under 80 mph–that people standing below could potentially catch them, though doing so would likely hurt their hands. And as for the person who dropped the penny, their clean criminal records would be pretty banged up as well (since dropping anything from the Empire State Building is highly illegal).

A penny would only reach a high enough speed to cause damage below if it were falling in a vacuum, or staying perfectly on edge all the way down–a virtual impossibility, unless you customize your coin by adding weight and stabilizing fins.

What if you dropped a baseball off the Empire State Building?

Don't try it. The ball's extra weight and shape would do some real damage. Not that it would necessarily kill someone below, but getting hit by a falling baseball sure wouldn't do them any good.

There is, in fact, some data to back this up. In 1938, for a publicity stunt, some Cleveland Indians caught baseballs that had been lobbed off a skyscraper. In response, the publicity man for the San Francisco Seals pro baseball team came up with an idea to go them one better. He figured that catching a baseball tossed out of the Goodyear Blimp would set a world's record and make a great season-opening stunt for the San Francisco team.

Not that he was completely unaware of the dangers. Figuring that extra padding and paraphernalia would lessen the impact, he recruited Seals catcher Joe Sprinz to do the honors while in full regalia.

At the appointed time, the blimp lumbered into position about 800 feet overhead. After much fanfare and ballyhoo, a player in the blimp dropped a ball and watched it disappear into the field below. It missed its mark and landed in the bleachers, cracking a board as it hit. A second ball at least landed within the playing field, landing outside the baseline and imbedding itself in the dirt. A lesser man might've quit, but Sprinz was determined to see the stunt through. Finally the blimp got into perfect position, and a third ball came rifling from the sky. Joe positioned himself underneath it and waited while the speck rapidly grew in size as it came screaming toward his outstretched glove. It hit the glove squarely and with a tremendous thwack, but it didn't stay there. Before Sprinz could close his hand and capture the ball, it bounced up and hit him in the face, splitting his lip, breaking his upper jaw, and knocking out several of his teeth.

The ball then dropped to the ground, and so did Joe. Still, he was triumphant as the guy who had attempted the highest catch ever recorded.

Over the years, other players have followed suit, but with

better preparation, lower heights, and no significant injuries. Baseballs have been caught from tall buildings, hovering helicopters, and the Washington Monument—all from below a height of 700 feet.

Catching a falling baseball requires quick reflexes, luck, and being prepared for terminal velocity—a term that, despite its ominous sound, doesn't mean "the speed that will kill you" but rather "the fastest speed an object can attain in a free fall." The terminal velocity of a baseball is only about 95 miles an hour. That's about the same speed a ball travels when thrown by a good fastball pitcher. In other words, a ball that's careening from a great height might be caught by a professional, but it can be dangerously unpredicable and potentially lethal if it hits the person's head instead of the glove. Fastballs can be lethal as well, which is why batters wear helmets.

How does the terminal velocity of other common objects compare with that of a baseball?

According to scientists who have tested such things, here are the approximate terminal speeds for various common objects. As you can see, variables include weight and wind resistance.

> Raindrop: 15–20 mph
> Ping-Pong ball: 20 mph
> Golf ball: 90 mph
> Baseball: 95 mph
> Person spread-eagled: 125 mph
> Person balled up: 200 mph
> 30-caliber bullet: 200 mph
> Bowling ball: 350 mph
> Space station falling from orbit: 900 mph

Hair & Dimples

How much does shaving off body hair help swimmers and bicyclists?

There's controversy about the effects shaving has on swimmers. A few studies using a small number of swimmers have shown some improvement in performance after shaving. So that may be the answer.

Skeptics, though, say the studies are flawed. Since most swimmers believe that shaving will help them go faster, they say it's not surprising that this expectation could psychologically affect their performance. The best studies would use a double-blind model in which neither researchers nor subjects would know if they had been shaved or not. They believe hair would make little or no difference, or even that hair might create turbulence like the dimples on golf balls (see below) to actually help a swimmer slip more easily through the water. "Otters and seals swim quite well, and they have lots of hair!" points out one antishaving swimmer.

On the other side, freshly shaved swimmers say their skin is more sensitive, so they have a better feel for their watery surroundings. They feel like they have more control over their movements and propulsion.

So the jury is still out for swimmers. Bikers, however, are likely just fooling themselves if they think shaving's going to help their time—sort of like the race car drivers who believe a clean car will travel faster than a dirty one. Smooth legs in biking offer more disadvantages—chafing, sunburn, infection from sweat dripping into the little nicks and cuts that inevitably result from shaving—than any perceived advantage, like speed. Also, there is absolutely nothing to indicate that a scrape from the road on a shaved leg is better than one on a hairy leg (although some bikers say that they like not having hair when they want to change Band-Aids). Hair can actually help act as an aid in clotting and prevent small dirt particles from getting into a scrape.

Who in the world thought of putting dimples on golf balls?

It was golfers themselves who discovered that golf balls with nicks, dents, and scrapes flew farther than the round, smooth balls that were originally used in the modern game of golf. The effects were so dramatic that they began aging their new balls as quickly as possible by pounding and scraping them. Eventually, the manufacturers recognized that they could sell more balls if they made the balls with dents already in them.

But what makes dimpled balls fly farther? Any object that's moving through the atmosphere has to push air out of its way. The air pushes back, producing friction and slowing the object down. Dimples help reduce the drag by creating a layer of turbulence around the ball—little swirling pockets of gusty air molecules that roll snugly along the outside of the ball's surface. Without dimples, the air would roll off the sides of the ball; the dimples part the air more neatly and force it to roll off the *back* instead of the *sides.* Besides reducing friction, this creates a back-pressure that helps propel the ball even farther.

But wait, there's more. There's also backspin to consider. The faces of all golf clubs—except for the putters—are slanted so that the ball gets launched into a backward spin as it flies through the air. This spinning motion pulls air quickly along the top of the ball, reducing air pressure in the same way an airplane wing does, lifting the ball and keeping it aloft for substantially longer than a smooth ball. (If you can hit the dimpled ball 260 yards, for example, you'll find that the same shot will send a smooth ball only 120 yards.)

How many dimples are on a golf ball?

The average ball has 415, with a range between 360 and 523 as the most popular.

Aural Fixation

What arrives first—a bullet or the sound of the gun firing?

"You never hear the one that hits you" was the fatalistic saying on the front trenches of many wars (not to mention the song of the same name by the rock group Stiff Little Fingers). Well, it turns out that the soldiers and the punk rockers had their physics down pat—a bullet travels faster than the speed of sound and will arrive at its destination before the report of the gun gets there.

How does a seashell sound like the ocean?

What you're hearing is part of the constant ambient sound that we don't often register because it always surrounds us. The shell reflects and focuses higher frequencies back to our ears. The reason is that high-pitched sounds have short wavelengths, so they tend to bounce when they hit a hard surface. The longer wavelengths of lower sounds tend to go through surfaces (which is why music through a wall or car window sounds so "boomy"). After bouncing around the shell's inside walls, the higher tones travel back to the ear randomly, creating a whooshing, echoey sound. The fact that a shell sort of sounds like rushing water is just a lucky coincidence. (To other listeners, it might just as easily sound like wintry winds, ventilator fans, or heavy traffic in the rain.)

Of course, there's nothing special about seashells doing this. You can do the same thing (in stereo yet!) with a coffee mug cupped over each ear.

Not surprisingly, if you're actually standing on the beach—a fairly noisy place—the "ocean" sound in the shell will be much louder than it would be in your own quiet kitchen. The same is true if you were standing in a crowded lobby of a building: the louder the environment, the louder the "ocean" sound.

If a tree falls in the forest with no one around to hear it, does it still make a sound?

Oh, no, you don't. This is one of those pointlessly endless questions with an answer that depends on how you define sound. If you define sound as "vibrations within a certain frequency," then the answer is yes. If you define it as "something perceived by an ear," then the answer is no.

You Don't Have to Be an Einstein

Is Einstein's brain still around, or was it buried with him?

After *New Jersey Monthly* journalist Steven Levy asked this question back in the mid-1970s, he tracked Einstein's brain down to a shelf in the study of Dr. Thomas Harvey—the doctor in Wichita, Kansas, who performed Einstein's autopsy in 1955. The brain was separated into lobes and kept in two mason jars inside a cardboard box marked with the label "Costa Cider."

Last we checked, Harvey still keeps the brain in a jar in his office, except for when he takes it on the occasional outing. For example, Harvey schlepped the brain cross-country to visit Einstein's granddaughter in 1997, reuniting the two generations after Einstein's death.

Where can I see pictures of Einstein's brain? Ask

Do geniuses have bigger brains than everyone else?

No, size doesn't seem to matter that much. During early tests on Albert Einstein's brain, for example, scientists found no significant difference between the size of his brain and those of people with normal intelligence. That said, there does seem to be evidence that genius brains may be different from average people's brains. Although each person's brain is as unique as his or her face, there are general similarities in shape and structures. Einstein's brain seemed to be lacking a specific wrinkle that the rest of us possess. That wouldn't be a big deal, except that without this wrinkle, Einstein's inferior parietal lobes–the part of the brain connected with mathematics, reasoning, and imagery–had more room to grow and are about 15 percent bigger than normal. Also, scientists speculate that the lack of the crease might make that part of the brain more efficient at communicating within itself. Whether this was why Einstein was so brilliant, no one really knows for sure. More research is definitely needed.

Can you explain the theory of relativity to me? **Ask**

C'mon Baby, Light My Fire

How many people have died from spontaneous combustion?

Well, let us answer your question as you asked it and then answer what we think you're really asking. Spontaneous combustion happens when heat-generating chemical reactions occur in an enclosed place. Usually heat from oily rags or mouldering vegetation escapes into the surrounding atmosphere, but if it's stored in a place with poor ventilation, the heat can't escape until it catches ablaze. Since there have been numerous cases of spontaneous combustion in warehouses and on farms, we know that many people have been killed as a result of these fires. However, it's hard to say for sure the cause of many fires, so attempting to quantify the numbers of human dead is difficult.

That's the answer to your question. However, we suspect you *might* have been trying to ask the following question, and so we'll ask it ourselves:

How many people have died from spontaneous human combustion?

Ah, yes, that mysterious phenomenon that people like to scare themselves with around campfires and on supernatural Web sites.

The stories have been around for centuries in legend, temperance tracts, and fiction. In *Jacob Faithful* (1834), novelist Frederick Marryat wrote about a character's disreputable mother who "perished in that very peculiar and dreadful manner, which does sometimes occur to those who indulge in an immoderate use of spirituous liquors. . . . She perished from what is termed spontaneous combustion, an inflammation of the gases generated from the spirits absorbed into the system."

In *Bleak House* (1853) Charles Dickens included a gruesome description of another drunken fellow's fiery demise: "There is a smouldering, suffocating vapour in the room and a dark, greasy coating on the walls and ceiling. . . . Here is a small burnt patch of flooring; here is the tinder from a little bundle of burnt paper, but not so light as usual, seeming to be steeped in something; and here is–is it the cinder of a small charred and broken log of wood sprinkled with white ashes, or is it coal? Oh, horror, he is here! Call the death by any name, it is inborn, inbred, engendered in the corrupted humours of the vicious body itself–Spontaneous Combustion, and none other of all the deaths that can be died." Herman Melville also recounted a similar incident in *Redburn* (1849) in which a drunken sailor spontaneously combusts before his shipmates' eyes.

These historic spontaneous combustion stories contained the recurring theory that a buildup of alcohol was responsible for the victim going up in flames. (The temperance movement often played up this angle in its literature.) Strangely enough, that's a recurring theme in modern stories as well–quite often the purported victims of spontaneous combustion are heavy drinkers (or users of sedatives), smokers, overweight, and elderly. They're found thoroughly burned, sometimes with just a small part of their body, like a foot, strangely unscathed.

And yet, despite the stories, exhaustive research indicates that spontaneous human combustion does not exist. One scientist went through 200 mysterious cases and came up with plausible reasons for all of them. Another, testing the long-standing theory of human body fat melting and absorbing into clothing, used a pig corpse dressed in nightclothes and managed to replicate the effect, common to most "spontaneous combustion" stories, of burning with a hot enough fire to consume flesh and bones but not nearby combustibles.

Here's a model that works for a large number of the cases: A smoker becomes unconscious from alcohol, sleeping pills, or a heart attack, and proceeds to drop a cigarette onto himself. It begins smoldering in clothes and eventually catches fire. The fat of the person, still deeply unconscious, melts and becomes absorbed into the clothes like wax from a candle. This wicking keeps the clothes from burning away, yet the fire burns with a continuous, hot, greasy flame. Like a candle, the flame is controlled and the heat largely flows upward toward the ceiling, covering it with a greasy soot but keeping the fire from spreading.

So the answer to the question is "Zero," as far as anyone can tell. No one has ever officially died of spontaneous human combustion.

Many Are Cold but Few Are Frozen

When was the thermometer invented?

The invention of the thermoscope—a device used to measure body temperature—is usually credited to the physician Santorio Santorio in 1612. Many argue that knowledgeable men of the day, including Galileo, collectively invented the device; however, Santorio was the first to actually add a number system to gauge changes in temperature.

Who invented the metric model for temperature?

Swede Anders Celsius did in 1742. But his version paradoxically had water boiling at 0 °C and freezing at 100 °C. Another Swede by the name of Carolus Linnaeus (who incidentally was also the guy who came up with the genus/species method of naming plants and animals) reversed these numbers and tinkered with the system a little. Despite Linnaeus's contributions, the scale was eventually named to honor Celsius. And thank goodness, it's much easier to spell.

How can I make a homemade thermometer?

How much faster will cold water freeze than hot water?

In theory, water that has been boiled and cooled will freeze faster than plain cold tap water. Boiling removes the air bubbles that tend to slow down the freezing process, and it also reduces the amount of liquid. However, in your kitchen or ours, the time difference is so minuscule, it wouldn't be noticeable. So don't go boiling all your water just to save time.

Elemental, My Dear Dmitri

Who invented the periodic table?

Russian scientist Dmitri Ivanovich Mendeleyev in 1869. In gratitude, scientists named element 101 after him: mendelevium.

Where can I find the Tom Lehrer song in which he sings all the elements?

Iodine Says "Aye," but Nobelium Says . . .

Can you tell me the symbol for the element nobelium?

No. (That's the symbol.)

Daddy Sang Base

Why are "noble" metals called that?

Noble, in this case, has nothing to do with royalty. It stands for metals like gold, silver, platinum, and mercury that are resistant to reactions with other things. They don't oxidize or rust quickly like base metals do. Base metals include copper, aluminum, lead, tin, nickel, and zinc.

Which elements are considered "rare," or in short supply?

Elements 57 through 71 plus the elements scandium (21), yttrium (39), and thorium (90) are all called *rare earth* elements. When many of these elements were discovered, the process of extracting them was difficult and costly. Scientists also believed they weren't largely available in Earth's crust. This is how they came to be called "rare." In actuality, most are fairly abundant, and with extraction so much easier these days, the only thing rare about them is their designation.

You, You, You Are the One, One, One

How many palindromes are found in the periodic table?
A palindrome, of course, is a word, phrase, or number sequence that can be read the same way forward and backward. An example would be the number 1234321 or the phrase "Egad, a base tone denotes a bad age." Based on that, then, there are no palindromes in the names of the elements. Of course, the atomic numbers make several palindromes (e.g., all of the single-digit numbers plus 11, 22, 33, etc.) but that's really boring. You can find quite a few element abbreviations with a single letter as well—technically making them palindromes: H (hydrogen), B (boron), C (carbon), N (nitrogen), O (oxygen), F (fluorine), P (phosphorus), S (sulfur), K (potassium), V (vanadium), Y (yttrium), I (iodine), W (tungsten), and U (uranium). Boy . . . ain't that fascinating.

All right, then. The only palindrome in the periodic table that even approaches cool has got to be unununium. No, the name itself doesn't work (though it almost looks like it should). However, its symbol is UUU. But wait, there's more: Unununium's atomic number is 111. You can't get much better than that. Trust us—we just spent more hours looking at the periodic table than we ever thought possible.

Poop Pourri

What's in animal dung that makes it flammable?
The undigested roughage—grasses, hay, and other vegetation—is what burns. Most people who use dried dung cakes for fuel add more dried hay to the fire to keep it burning longer.

Burning manure produces a lot of smoke, though, putting those in the home at risk for lung problems. But fermented dung produces a lot of methane gas, which makes a relatively cheap and unpolluting fuel.

Where can I see a list of the ingredients in poop?

Ask

Accounting for Inflation

Is air weightless?

Just because something's invisible doesn't mean it's weightless. As a matter of fact, all gases have mass, and therefore they have weight. Air, a combination of gases, is fairly heavy. Good thing, too–the weight of the air around us and the force it exerts on our bodies pushes on us, and that is what keeps us from exploding. The pressure inside our bodies, in turn, keeps the air pressure from squishing us flat. Air weighs about .07 pounds per cubic foot. And believe it or not, each square inch of our bodies is under about 14.7 pounds of air pressure.

Well, how about helium, then? Is a helium balloon weightless?

Helium is the second lightest gas. Hydrogen's the lightest. Both gases have mass and therefore weight. It just so happens that they weigh less than air. Because of that, the heavier air sinks below them, pushing them upward–much in the same way that water flows below Styrofoam and pushes it toward the surface. If released into the air, helium and hydrogen atoms will rise up toward the top of the earth's atmosphere; if captured inside a balloon, the gas will do the same.

The Hunting of the Quark

Where did the term quark come from?

Quark is a nonsense word, coined by the writer James Joyce in *Finnegans Wake.* In the book, it may have referred to a cheer or the call of a gull, but these things are never completely clear when referencing the writings of the impossibly inscrutable Irish writer.

In science, of course, a quark is nearly as inscrutable. It's believed to be a tiny particle within the nucleus of an atom that binds together with other quarks to eventually make up protons. There are six known types of quarks, referred to as (ready for this?) up, down, charm, strange, bottom, and top. You can imagine what kind of surreal Abbott and Costello routine could've come from that ("Hey, Abbott, what's up?" "Down." "Up is down? That's strange." "No, no, that's charm. . . .").

Confusing? Yeah, well, that's not surprising, since they are theoretical (nobody's ever actually seen one) and so tiny they're hard for most of us to comprehend. Simply put, they're believed to be the cosmic LEGOs that make up all matter.

The origin of the particle's name is rather creative, and nearly as confusing as the particle itself. In 1963 a brilliant, Nobel Prize–winning theoretical physicist named Murray Gell-Mann offhandedly gave the particle the whimsical name *kwork,* in much the same way you or I would use the term *gizmo* or *thingamabob.* Although Gell-Mann was also a James Joyce fan, he had no recollection of having ever run across the word *quark* in his reading. In his 1995 book *The Quark and the Jaguar,* Gell-Mann tells the story behind the naming:

> In 1963, when I assigned the name "quark" to the fundamental constituents of the nucleon, I had the sound first, without the spelling, which could have been "kwork." Then, in one of my occasional perusals of *Finnegan's Wake,* by James Joyce, I came across the word "quark" in the phrase "Three quarks for Muster Mark." Since "quark" (meaning, for one thing, the cry of a gull) was clearly intended to rhyme with "Mark," as well as "bark" and other such words, I had to find an excuse to pronounce it as "kwork." But the book represents the dreams of a publican named Humphrey Chimpden Earwicker. Words in the text are typically drawn from several sources at once, like the "portmanteau words" in *Through the Looking Glass.* From time to time, phrases occur in the book that are partially determined by calls for drinks at the bar. I argued, therefore, that perhaps one of the multiple sources of the cry "Three quarks for Muster Mark" might be "Three quarts for Mister

Mark," in which case the pronunciation "kwork" would not be totally unjustified. In any case, the number three fitted perfectly the way quarks occur in nature.

Just Count One-Mississippi, Two-Mississippi . . .

My brother dared me to count to a billion once, but you can't do that in your lifetime, can you?

Sure you could. Theoretically, anyway. There are 86,400 seconds in every day. If you gave yourself a second per number, this means you could count to a million in just over 11.5 days, and a billion in about 11,574.5 days–or about 31 years and 252 days. Of course, the figures above don't factor in sleeping, eating, or living a normal life. So if you'd like to count only half the day and take the rest of the time off, you'd have to allow for more than 23 days for a million, and 63 years and 137 days to count to a billion.

If you lived in Great Britain, you'd have an even harder time: An American billion is only a thousand million (1,000,000,000), but a British billion is a million million, or what Americans would call a trillion (1,000,000,000,000). To get to that number, you could start in this lifetime, but you'd have to leave the finish to one of your descendants: counting a number per second for twelve hours a day, seven days a week, would get you to a million million in 63,376 years and 65 days. Better get started.

Where can I see algorithm animation movies online?

Ask

Buzz Light-year, for One

What's beyond infinity?

In our number system, nothing. Infinity doesn't really exist. It is simply a concept, not a number or a thing in itself. If there's anything "beyond" infinity as we understand it, it's part of infinity. So, the answer is that if there's anything beyond infinity, it's just more infinity.

"May I Have a Round Mnemonics Pi, Jeeves?"

I seem to remember a phrase or rhyme to help remember the numbers of pi. Do you know what it is?

There are several mnemonic ways to memorize pi. Most involve remembering a sentence or rhyme and counting the letters within each word. For example, count the letters of each word of the following phrase: "May I have a large container of coffee? Thank you." "May" has three letters, "I" has one, "have" has four letters, and so on. Put them side by side, add a decimal point after the first digit, and eventually you'll get 3.141592653, or pi to ten digits.

Better still is to memorize a rhyme that will get you 31 digits, which is many more than you'll ever need to get an accurate calculation for most purposes. Here's an old one by "Anonymous":

> Now I will a rhyme construct,
> By chosen words the young instruct,
> Cunningly devised endeavors,
> See it and remember ever
> Widths in circle here you see
> Sketched out in strange obscurity.

We'd be remiss to not mention that National Pi Day is officially held on—you guessed it—March 14, at 1:59 (you know, 3/14 1:59).

Which has more pizza, a ten-inch pizza or two seven-inch pizzas?

Next time you hear someone complain about never using the math you learned in school in real life, remember this question.

You'd think the two smaller pies would have more, but that would be wrong. You just need some pi with your pizza.

Remember the formula? $A = \pi R^2$. (The "pi R squared" thing works in this case even though the pies are round.) The area of a circle is the same as half the diameter (5 inches for the ten-inch and 3.5 inches for the seven-inch) multiplied by itself (25 and 12.25) times 3.1417 (pi). So a ten-inch pizza is the better deal. It has about 78.5 square inches of cheesy goodness, while two seven-inch pizzas have only about 77.

Thanks for Nothing

Didn't the Arabs invent the Arabic number system?

No, the Hindus of India did, around A.D. 600. It was, however, an esteemed Muslim mathematician, al-Khwarizmi, who introduced the system to the Western world when his treatise on mathematics was translated into Latin in the tenth century. Which is why Westerners mistakenly attributed it to the Arabs.

The Hindu number symbols we use today originated as early as the 200s B.C., with one exception: the zero. That didn't appear until about A.D. 600. Its name probably comes from a garbled translation of the Hindu word *sunya,* which means "empty."

Is there a Year Zero in our calendar?

Nope. The year before Christ was theoretically born is referred to as 1 B.C.; the year he was born is counted as A.D. 1, with no Year Zero in between. And if Jesus were born in A.D. 1, on December 25 (which he wasn't), it would be mighty close to A.D. 2. That just doesn't make much sense.

Math Hysteria

Who invented geometry?

It may very well have been some sadistic teacher in ancient Egypt. Most scholars believe that the Egyptians were the first to make extensive use of geometry, what with measuring land, building those pyramids, and all.

The truth is that geometry as a science has very obscure origins. Geometry in general, though, has been around as long as humanity has been able to reason. The concept would be mighty hard to miss in this world full of shapes and angles.

We do know that the name comes from two Greek words meaning "earth" and "measure."

But here's one part of geometry that you can pin on someone: Remember coordinate plane geometry? Here's a quick reminder: on a grid sits the X and Y coordinate, two lines that intersect, with numbers heading out in two different directions. By tracing your finger to an object or point, you can find the coordinate points for anything on the grid.

Anyway, for those long hours spent sketching out parabolas, you can blame philosopher and mathematician Rene Descartes (1596–1650). The story goes that while lying in bed, he noticed a fly walking on the ceiling. Descartes began idly wondering if he could create a system to chart the bug's path. Coordinate geometry was his solution, much to the chagrin of math students ever since.

What does the name calculus *actually mean?*

The Pythagoreans—ancient Greek followers of the mathematician Pythagoras—were a brilliant lot. They came up with the mathematical truism, The square of the length of the hypotenuse is equal to the sum of the squares of the lengths of the other two sides. They also mathematically figured out the correct shape of Earth. In working out mathematical equations, the Pythagoreans practiced simplicity, using little rocks to represent numbers. Thus, the word *calculus* means "pebbles" in Latin.

Quicker than a Wink

So how long is a jiffy?

Webster's Unabridged dates the term back to the early eighteenth century. It was used then, much as it is now, to mean "instant," as in, "This chapter will be done in a jiffy!" It's also been used as a computer term. Depending on which source you ask, it's either one-sixtieth of a second or one one-hundredth of a second.

Bathrooms

It's perhaps the smallest room in the house, and yet significant enough to have a chapter of its own. Clean water comes flowing in, unspeakable filth gets transported out, and in between there's a place of blessed solitude.

Water Flowing Underground

What do cities do to purify water?

Water from lakes, rivers, and reservoirs isn't usually clean enough for our drinking glasses, so before it gets to our tap, it goes through a water treatment plant.

First the water runs through large filters that rid it of large objects like sticks, leaves, dead animals, trash, and so on. From there, the plant injects alum—a salt that gets sticky when stirred with water. It attracts grit, sand, and dirt. Clumps of these substances, called *floc,* fall to the bottom of the tank, where they're removed.

The cleaner water is forced through more and finer filters as well as through activated charcoal, which further absorbs small particles. Finally, the water is spritzed with a small amount of chlorine—to kill viruses, bacteria, and certain harmful amoebas—and fluoride to strengthen teeth. After this, the water's sent flowing underground toward your water taps.

Down the Tubes

If our sink drains lead straight to the sewers, how come they don't smell that bad?

Credit the "stink trap," which pretty much does what its name implies. It stops the stench of the sewer from rising up and bothering the inhabitants of your house.

How does the stink trap work? Oh, it's a marvel of primitive technology, but it took humans a remarkably long time to figure it out. (In fact, President Chester Arthur reportedly fled the White House in 1882 in fear and revulsion when a sewer sent gas up through the sinks, baths, and toilets. If only he'd had stink traps.) The stink trap is that loop in the pipe that's under your sink. In the old days, the pipe was shaped like an S, but nowadays, it's shaped more like a P facing downward. (Ironically, toilets still have an S-trap when a "P"-trap would be a much more appropriate name.)

How does the loop shape keep the smell out? Water gets trapped in the lower loop of the pipe, creating a seal that doesn't allow gases to pass either way. That keeps sewer gases out of your house (and, for that matter, your house gases out of the sewers).

Have alligators ever lived in sewers?

No, that's a myth. Goldfish won't survive there either, no matter what your older brother might've told you before he flushed yours.

Where does it go when I flush the toilet? Into some river somewhere?

Not anymore. At least not right away, unless it rains and the extra liquid seeping into pipes overwhelms the sewage system. Here's what happens:

1. If you have a septic tank buried in your backyard somewhere, the waste and water go there. The solid wastes sink to the bottom of the tank and eventually get reduced in bulk by anaerobic bacteria; the liquid gets spread underground to soak into what's called the "drainfield"–in other words, your backyard.
2. If you're connected to a sewage system, your toilet waste goes on a little ride. After coursing through a pipe inside your walls and under your floor, the waste ends up in a sewer pipe underneath your street. If you want to know exactly where, look for the manholes–they're the way crews get into the sewer if a problem arises.

After traveling through a succession of progressively larger pipes, your waste will end up in a sewage treatment plant, where it joins millions of gallons of other sewage. There it goes through several stages:

1. Straining out solid stuff like dead goldfish, condoms, menstrual products, dental floss, and paper.
2. Letting grit settle out. "Grit" refers to anything that sinks, like sand, gravel, coffee grounds, eggshells, and much of the stuff that people grind up in their kitchen garbage disposals. The method is simple: in a tank, air jets bubble up scum and other lighter stuff and allow heavier stuff to hit bottom. A giant screw in the bottom of the tank continuously pushes the grit out of the tank.
3. Settling out the sludge. In the next pool, the sewage sits for about two hours, allowing sludge solids to settle to the bottom of the tank. What look like extra-long bicycle chains glide bars across the top of the tank to skim off scum. A similar system is working unseen at the bottom of the tank, scraping sludge to a low spot on one end, where a pump sucks it out and sends it for digestion, dewatering,

and disposal into landfills, farmland, or fertilizer bags in your local nursery.

4. In the next tank, the wastewater gets dosed with tons of oxygen—about 140 tons a day. While a huge blender blade stirs the oxygenated mixture, the plant adds "activated sludge" enriched with microorganisms. This sludge acts like yeast in bread dough, with its organisms multiplying like microscopic rabbits in the oxygen-rich water and gobbling up most of the remaining pollutants.

5. After its two hours in the oxygenation tank, the water moves into a big round pool called a secondary clarifier. As the water flows through, giant scrapers rotate constantly along the bottom, moving solids out of the tank. Most of those solids are the microorganisms from the previous step, which are shipped back to the oxygenation tank to culture the next batch of wastewater.

The water—now clear and smelling like water—is doused with chlorine to kill pathogens, then doused with sulfur dioxide to neutralize the chlorine's negative effects on wildlife. *Then,* finally, it's dumped into a public waterway.

Swirls Apart

Is it true that the water in toilets and bathtubs spins clockwise in the Northern Hemisphere and counterclockwise in the Southern?

No, it's a myth. There is something that's called the Coriolis force, which affects huge bodies of water and tends to make them want to spin in response to Earth's rotation. But what works with huge oceans doesn't work on the much smaller scale of bathroom fixtures. Good thing, too—otherwise you'd have to worry about your toilet overflowing twice a day at high tide.

Who or what is a Coriolis?

Gaspard-Gustave de Coriolis (1792–1843) was a French scientist who not only discovered the Coriolis force that bears his name, but also introduced the concepts of "work" and "kinetic energy" to physics. He also tried to popularize a new word he coined, "dynamode," which he defined as 1,000 kilogram-meters of work (in other words, the amount of energy required to move one kilogram a thousand meters, or a thousand kilograms one meter). But the term died a quick and merciful death.

Next to Godliness

Is there really a medical reason for bathing?

If you don't mind stinking, no. Your skin doesn't care if it's clean or dirty. In fact, excessive bathing can do some harm—it can irritate the skin, and dry it out in older people. Throughout history, there have been some people who have prided themselves on their ability to count their total number of baths on the fingers of one hand, and sometimes on no fingers at all.

However, you will smell bad. While bacteria efficiently take the place of soap and water, eating the skin cells and sweat that you'd normally wash off, the substances they excrete are one of the major causes of people's stinky smell.

How often do I have to bathe to stay clean?

It's a judgment call that depends on what you mean by clean. Advertisements of soap and deodorant companies imply that you should bathe daily or even more frequently. In some parts of the world, it's common practice to bathe only once a week or less, with regular rinsing of the odiferous body parts considered otherwise sufficient.

It may simply come down to local custom. If you're with people who bathe once a month, they presumably won't be offended by your stench if you do the same. However, if the people around you bathe every day, you may feel like you reek if you go two or three days without a shower.

What do we say? If your bathing regimen removes the dirt and dead skin from your body, and you're happy with the way you smell, then stick with it. We've been conditioned by years of marketing ploys to worry unduly about smelling bad. In fact, the term *B.O.* was coined by an advertising man in 1919 in order to sell Odo-Ro-No, one of the first deodorants. Wrote a wise marketer long ago: "Advertising helps to keep the masses dissatisfied with their mode of life, discontented with ugly things around them. Satisfied customers are not as profitable as discontented ones."

Gray Grows the Mildew-o

How come mildew only grows in between the tiles on the shower wall?

It has to do with the porousness of the surface. Ceramic tile is glazed and therefore not porous like the grout between the tiles. Because it retains water, the grout creates a fertile field in which mildew can grow. The truth is, though, if you don't use cleaners that are abrasive and have bleaching agents, you might not have a mildew problem at all. Harsh cleansers eat away at the grout, increasing its tendency to retain water and allowing the mildew to further seep into the spaces between the tiles.

In an emergency, if the water pipes are broken and we need water to drink, how much bleach should I use to purify the water we can find?

Well, first of all, try using the several gallons of water in the reservoir tank of the toilet (the squarish thing at your back when you sit down on the seat—not the bowl). It's fresh and clean enough to drink, unless you're a little squeamish about drinking out of any part of a toilet. (Why should you be? Your dog probably does it all the time.) If so, you can purify the water with a tiny amount of bleach.

First the warnings: Undiluted bleach is poisonous, and should not be drunk, or spilled on your skin or eyes, and its fumes should not be inhaled for any significant amount of time. Only bleach labeled "liquid chlorine bleach" should be used. It should be 5.25 percent hypochlorite—read the label carefully—and should not have scent or other cleaning agents added to it.

Follow these mixing guidelines to purify water in an emergency: For every gallon of clear water, add 8 drops of liquid chlorine bleach. For cloudy water from more dubious sources, double that amount to 16 drops. Stir the mixture, leave it in an open container, and wait for at least half an hour before drinking so most of the bleach can evaporate off.

Keep in mind that bleach will kill many of the pathogens that can make you sick, but it will not neutralize chemical poisons or pollutants.

Rub-a-Dub-Dub

Why does the hot water tap in my shower start out fast, but then slow down to a trickle?

It sounds like you have a defective washer, that rubber disk within the faucet that helps stop the water from flowing when you turn off the tap.

Although the washers in both faucets can erode with time, the hot water washer is likely to go bad much more quickly. What happens is that hot water expands the rubber washer, and then as the washer cools, it shrinks again. This constant change wears out the hot water washer faster than the one in your cold water faucet. The deteriorating washer can produce a number of results, from leaky, dripping faucets to a high-pitched squeal when you turn on the water.

In your case, as the defective washer gets hot, it swells and partially blocks the water flow, reducing output to almost nothing. In other words, replace it.

I take long showers. Would I save more water taking a bath?

Only you can test this one out. Plug your tub while showering, and see how high the water level is when you're done. In general, a shower usually uses 10 to 15 gallons of water, while a bath usually takes about 15 to 25 gallons. But conduct your own measuring experiments to get more personal results.

Good Clean Fundamental Science

Could you tell me the story of Archimedes' inspiration in the bathtub?

Better yet, we can give you an account from the first century B.C., written by Marcus Vitruvius Pollio, a Roman architect, engineer, and writer:

> In the case of Archimedes, although he made many wonderful discoveries of diverse kinds, yet of them all, the following, which I shall relate, seems to have been the result of a boundless ingenuity. Hiero, after gaining the royal power in Syracuse, resolved, as a consequence of

his successful exploits, to place in a certain temple a golden crown, which he had vowed to the immortal gods. He contracted for its making at a fixed price, and weighed out a precise amount of gold to the contractor. At the appointed time the latter delivered to the king's satisfaction an exquisitely finished piece of handiwork, and it appeared that in weight the crown corresponded precisely to what the gold had weighed.

But afterwards a charge was made that gold had been abstracted and an equivalent weight of silver had been added in the manufacture of the crown. Hiero, thinking it an outrage that he had been tricked, and yet not knowing how to detect the theft, requested Archimedes to consider the matter. The latter, while the case was still on his mind, happened to go to the bath, and on getting into a tub, observed that the more his body sank into it the more water ran out over the tub. As this pointed out the way to explain the case in question, he jumped out of the tub and rushed home naked, crying with a loud voice that he had found what he was seeking; for as he ran he shouted repeatedly in Greek, "Eureka [I have found it]."

Taking this as the beginning of his discovery, it is said that he made two masses of the same weight as the crown, one of gold and the other of silver. After making them, he filled a large vessel with water to the very brim, and dropped the mass of silver into it. As much water ran out as was equal in bulk to that of the silver sunk in the vessel. Then, taking out the mass, he poured back the lost quantity of water, using a pint measure, until it was level with the brim as it had been before. Thus he found the weight of silver corresponding to a definite quantity of water.

After this experiment, he likewise dropped the mass of gold into the full vessel and, on taking it out and measuring as before, found that not so much water was lost, but a smaller quantity: namely, as much less as a mass of gold lacks in bulk compared to a mass of silver of the same weight. Finally, filling the vessel again and dropping the crown itself into the same quantity of water, he found that more water ran over the crown than for the mass of gold of the same weight. Hence, reasoning from the fact that more water was lost in the case of the crown than in that of the mass, he detected the mixing of silver with the gold, and made the theft of the contractor perfectly clear.

It's a great story, written two centuries after Archimedes' death (he was killed by a Roman soldier during the sacking of Syracuse). It might even be true, although Chris Rorres, a mathematics professor at Drexel University, gives us some reason to doubt. "Suppose the dishonest goldsmith replaced 30 percent (300 grams) of the gold in the wreath by silver," he writes. "The difference in the level of water displaced by the wreath and the gold is 0.41 millimeters. This is much too small a difference to accurately measure the overflow from, considering the possible sources of error due to surface tension, water clinging to the gold upon removal, air bubbles being trapped in the lacy wreath, and so forth."

Be True to Your Teeth & They Won't Be False to You

How does fluoride prevent cavities?

As it turns out, in at least three ways. Science has known for a while that fluoride acts as a building block for teeth. Fluoride supplements, whether in drinking water or vitamins, help build stronger enamel in kids' teeth before they even emerge from the gums.

After the teeth emerge, fluoride, regularly applied, settles into the nooks and crannies in the teeth that are formed when acid begins breaking down the surface of the enamel (the beginnings of a cavity). The fluoride attracts calcium and other strengthening minerals to the weakened site. Repeated exposure to fluoride ends up forming a new layer of tooth that is harder and stronger than the original.

What's interesting is a more recent discovery about another role of fluoride in fighting tooth decay. It appears that fluoride residue in your mouth neutralizes the bacteria in plaque and keeps it from forming acidic, tooth-rotting sugars.

Ironically, it goes to show that you should probably drink fluoridated tap water instead of that bottled stuff you thought was somehow healthier. You should also use a fluoride toothpaste. And no matter what your mom may have told you, you should avoid rinsing out your mouth after brushing your teeth so that the fluoride can stay around and kill all of those harmful bacteria.

Lather, Rinse, Repeat

Why don't we use one soap for all needs? Why do we use shampoo?

There are good reasons for using each for specific purposes. Soap cleans by using two methods. First, the fat in soap seeks out and surrounds dirt. The other substances in soap cling to water and help to wash away the fat and dirt molecules.

But soap doesn't work well on hair. Because soap contains fat, it doesn't mix with water and leaves behind a film. On your hands and skin, you don't really notice this much, and it quickly gets sloughed away with dead skin. Your hair is another matter: the film weighs down the strands and makes them dull, dingy, and somewhat greasy-looking. It's just not aesthetically pleasing.

That's why shampoo is made of a detergent, not soap. Instead of having a two-part reaction, the dirt-seeking molecules in detergents are mixed with alcohol. These alcohol-besotted molecules mix thoroughly with water, and so rinse away completely without leaving a film. The downside of any product containing alcohol is that if it is used on skin, it tends to dry it out, so soap works best on skin, and detergent on hair.

Why are vitamins put in shampoo?

Marketing. Because they make the shampoo sound like it's more than just expensive glorified dish detergent. Although a lot of shampoo brands imply that the various vitamins they put in their products help make hair healthy and shiny, the truth is that vitamins can't do much of anything for your hair. The hair is already dead, and won't really absorb or react to anything you put on it.

It's true that vitamin deficiencies can affect hair, but in that case dull hair is likely going to be the least of your problems. If daily vitamins (the kind you swallow, of course) are taken regularly, they can help your whole body function better—

including the hair follicles that produce hair–but from a scientific standpoint, no nutrients you place on your hair or skin are going to have much benefit.

How long does it take to get my hands clean with a bar of soap?

From a bacterial standpoint, it takes about twenty seconds for the germs and dirt to be picked up and washed away.

What do they put in Ivory soap to make it float?

Air. They whip the soap, allowing air pockets to get into the mixture before it solidifies. As with a flotation device in a swimming pool, the air is what makes the soap buoyant.

Hair Today, Gone Tomorrow

How does Rogaine make your hair start growing again?

Actually, no one's really sure how it does it. What we do know is that it can't make hair suddenly start growing after years of not growing. Because of this, it's most effective on people who have just begun to notice a little bit of thinning. On most people, after about two weeks some hair starts to grow. But if the treatment is stopped, the hair stops, too.

Getting Around

How come sailboats can travel into the wind? What happens if lightning strikes an airplane? Are freeways really designed as emergency landing strips? Sometimes the farther one travels, the less one knows.

Waiting for Mr. Wright

My friend swears that the Wright brothers weren't the first people to fly in heavier-than-air craft. He's wrong, right?

Wrong. He's right, actually. People rode around in gliders before the Wright brothers were even born.

In 1849 British engineer George Cayley built a glider that he tested by flying it like a kite. It had enough lift to carry a small boy briefly into the air, but it still wasn't large or maneuverable enough to fly untethered. Finally, in 1853, Cayley built a larger glider and coaxed his coachman into climbing aboard it. The poor fellow flew 500 yards and into the history books. (He also flew into the unemployment line. Upon landing, he reportedly told Cayley, "I wish to give notice, sir–I was hired to drive, not to fly.")

Other pioneers followed Cayley's lead, including German engineer Otto Lilienthal, who regularly flew gliders made from waxed cloth and willow sticks, and of course Orville and Wilbur Wright, whose gliders advanced the knowledge of lift and maneuverability. In 1903, the Wrights took the next logical step–adding a small gasoline engine with a propeller to one of their gliding contraptions. On its first flight at Kill Devil Hill near Kitty Hawk, North Carolina, the *Flyer,* piloted by Orville, traveled about 120 feet at approximately 30 mph. The brothers flew three more flights that day, alternating as pilots. The longest flight, with Wilbur at the controls, was 852 feet and lasted 59 seconds. Not many people heard or read about the experiments at the time, and even fewer still imagined that the experimental craft would ever show much promise except as a curiosity.

Why didn't anybody else think of adding an engine to a glider before the Wrights?

A lot of people did think about it. For example, in 1843, William Samuel Henson wrote an article in *Mechanics Magazine* that proposed an "Aerial Steam Carriage" that would use a steam engine to drive a propeller in a fixed-wing aircraft. The problem wasn't a lack of imagination, but of the right technology. Steam engines were too heavy to do the job–it wasn't until auto builders perfected the smaller, lighter internal combustion engine that powered fixed-wing flight became possible.

You Can't See Me

Why can't the stealth bomber be picked up by radar?

Designers of the B-2 Spirit stealth bomber say that they used radar-absorbing surfaces and a weird shape to outsmart radar systems. However, the more complicated and more disturbing answer is that the much-ballyhooed stealth design works only against a certain class of radar, and that the planes can be detected by newer radar systems. As a result, the bombers now "work" only when surrounded by radar-jamming planes.

This is nothing new. The original "radar-proof" aircraft had a similar problem. The U-2 spy aircraft of the 1950s were essentially gliders made of plastic and plywood that were powered by a jet engine. They were designed to be small and light enough to cruise for long periods of time at very high altitudes (80,000 feet) that were beyond the range of antiaircraft fire. However, after several flights, another unexpected benefit showed up: the U-2s weren't being picked up by Soviet radar. For more than four years, U-2s remained virtually unchallenged in Soviet skies. That invulnerability, however, came to an end after the USSR upgraded both its radar and its antiaircraft weaponry, and in May 1960, the Russians shot one down. Pilot Gary Powers ejected and was captured, creating an embarrassing international incident for the United States.

And so it goes with new advances in weaponry: it doesn't take long before someone figures out how to beat it, sometimes accidentally. For example, the illusory invisibility of the stealth system took a big hit in Yugoslavia in 1999 when a bomber was shot down by Serbian troops using an antiquated, low-frequency radar system that wasn't fooled by its high-tech trickery. Now most new radar systems use a mix of high and low frequencies, rendering the stealth technology increasingly obsolete.

Slipping the Surly Bonds

Has anybody ever used a bunch of helium balloons to fly into the air?

It's an old childhood dream, isn't it? Getting together enough helium balloons to "slip the surly bonds of earth" and glide peacefully over the rooftops. Well, that was Larry Walters's idea as well. On July 2, 1982, the North Hollywood truck driver tied forty-five weather balloons to a lawn chair outfitted with a pellet gun, a CB radio, a giant bottle of soda, and water-filled milk cartons as ballast. Walters had hoped to slowly waft into the sky, but instead shot up to an altitude of 16,000 feet.

Forty-five minutes into the flight, still flying at this very high altitude and becoming increasingly numb from the cold and thin air, Walters began shooting at the balloons. In the process he ended up dropping the gun, but had luckily popped enough to start a slow descent. After a total of about two hours in the air, his balloons tangled in power lines, blacking out a Long Beach neighborhood for twenty minutes, but he was able to scramble unhurt from his dangling lawn chair. His stunt earned him an appearance on *Late Night with David Letterman* and a fine of $1,500 from the Federal Aviation Administration for reckless flying, failing to stay in contact with traffic controllers, and "piloting a civil aircraft for which there is not currently in effect an airworthiness certificate."

On Gossamer Wings

When was the first human-powered airplane flight?

The ancient myth of Daedalus and Icarus notwithstanding, it wasn't until 1977 that a human-powered airplane managed to go any significant distance. Called the *Gossamer Condor,* the plane was designed by a California-based engineer named Paul MacCready. MacCready built the plane in order to collect on a long-standing award that British industrialist Henry Kremer established in 1959, which awarded £50,000 that first year to the first substantial flight of a human-powered plane.

Despite measuring 30 feet long with a wingspan of 96 feet, the *Condor* weighed just 70 pounds. How did they make it so light? Its skeleton was made of aluminum tubes braced with piano wire. Balsa wood, corrugated cardboard, Styrofoam, and plastic sheeting made up the body and wings, and a bicycle crank and chain connected the power train to the propeller.

Strong-legged bicycle racer Bryan Allen piloted it. To win, the *Condor* had to pass over a ten-foot barrier at the start, fly in a figure eight around two pylons set half a mile apart, then pass again over the ten-foot barrier before landing. The plane traveled the 1.25-mile course in 7 minutes, 22.5 seconds, averaging just over 10 mph. For their troubles, MacCready and Allen split the prize, equivalent to $95,000.

Bolt from the Blue

What happens when lightning hits an airplane?

If you've got a fear of flying and an active imagination, your mind can generate vividly horrifying scenes: ball lightning rolling down the aisle, shorting out laptops and cell phones, electrocuting passengers, overcooking the chicken à la king, that sort of thing. It probably doesn't help to hear that lightning strikes airplanes more often than you might think (in fact, the mere presence of the airplane near highly charged clouds can trigger a lightning bolt).

The good news is that being in a plane struck by lightning isn't usually a problem. The airplane's skin is designed to conduct the electricity around the passenger compartment and discharge the energy back into the atmosphere. If you look, you can see the little rods of the "discharge wicks" on the rear edges of the plane's tail and wings.

That wasn't always the case, however. Regulators mandated various protective measures after lightning caused the crash of a 707 in 1963. Since then, aside from an occasional report of some minor damage to a plane's electrical systems, lightning hasn't been implicated in anything like what our imaginations can conjure up. So next time you're flying in a storm, blame the overcooked chicken dish on something else.

Where can I see a video of a plane being struck by lightning?

Let It Glide

If a large commercial jet loses its engines, can it glide to a landing, or will it just drop like a rock?

A lot of nervous fliers have come under the impression that passenger jets will pretty much fall out of the sky if all the engines fail. The truth is that large planes can glide about as well as small ones. (In fact, sometimes while landing a pilot will put the engines into idle and let the plane simply glide onto the runway.) That's the reassuring part. What makes gliding to earth more dangerous with a large commercial jet is that it goes faster and builds more momentum as it descends and therefore requires a longer emergency runway than a small plane does. Finding one of these quickly can be a challenge.

Into Thin Air

Flying from South America, we were told it was too hot for the plane to depart fully loaded. Is that possible?

It is. Increasing temperature or altitude can make a plane fly less efficiently by decreasing the overall density of air. The heat wouldn't make the plane unable to fly, but it could increase the distance the plane needs to lift off and reach a good altitude. That could be a problem when taking off on a short runway or toward tall obstacles like buildings or mountains.

Whiplash!

How do planes on aircraft carriers take off on such a short runway?

They use a technology that was perfected in the days of chariots and togas: the catapult. Under the deck of the carrier are two huge cylinders that measure more than 150 feet long. The cylinders, connected to a launcher, build up a great deal of pressure as they're filled with steam diverted from the ship's engines. When the pressure's released, the catapult can launch planes from 0 to 165 mph in 2 seconds. Landing is also a problem on such small real estate. A hook on the plane's tail grabs an arresting cable that stretches across the deck, bringing the plane to a screeching halt. All this jerking around makes us wonder if the navy should recruit more chiropractors.

Is it true that every fifth mile on the federal highway system is made straight so planes can land on it?

While planes occasionally make an emergency landing on a freeway, there isn't (and never has been) any such policy. "As with Dracula, it is very difficult to put a stake through the heart of this 'fact,' " writes Richard F. Weingroff, an information liaison specialist for the Federal Highway Administration's

Office of Infrastructure. "It's like the 'urban myths' that people nevertheless believe. Nevertheless, no law, regulation, or policy requires that one out of five miles of the Interstate Highway System must be straight."

A Good Reflection on You

When I tilt my car's rearview mirror, it deflects headlights. How does it do that?

Well, it's really quite clever in a low-tech sort of way. Picture a sheet of glass that isn't the same thickness all the way through, but instead is cut at an angle so its top is thicker than its bottom. If you put reflective silver on the back of it and mount it on your windshield, the back surface will reflect light normally during the day, just like a regular mirror.

In the daylight the clear front surface of the glass is also reflecting back a little light. Why? Because it's angled down slightly, so on close inspection, you might be able to see a little ghost image of your back seat. Normally, though, you don't notice it's there.

In night mode, however, the clear glass surface becomes what you use as your rearview mirror. With a slight tilt of the assembly, you've pointed the clear glass surface toward your back window and the glaring mirrored surface up toward the ceiling. Only about 4 percent of the light from the cars behind you gets reflected toward your eyes by the clear glass; the rest passes through to the mirrored glass, which reflects it harmlessly toward your ceiling.

Them's the Brakes

How much farther does it take to stop a car going 60 mph than one going 30?

You'd think it would take twice as far, but in fact it'll take about *four times* as far—roughly 180 feet at 60 mph, versus 45 feet at 30 mph. It's an illustration of the principle that as you increase speed, your momentum (and thus your stopping distance) increases exponentially.

Blech!

Why do people get motion sickness?

The theory we like best is that it's your body's reaction to perceptual dissonance. While riding in a car, you're in an enclosed area in which your eyes perceive that nothing nearby is moving, but your ears and the rest of your body feel like you indeed are moving. In an effort to solve that enigma, your mind starts your body puking, perhaps interpreting the inexplicable dizziness as an onslaught of poisoning. (Compare, for example, the body's reaction to a first experiment with tobacco or too much alcohol.)

Given that, the nonmedical cure for motion sickness involves making sure your eyes are in sync with your ears by looking out the window toward the horizon. Above all, avoid anything (reading, for example) that keeps your eyes focused inside the passenger compartment.

Yes, the whole theory sounds a little cockamamy, but the cure seems to work for most people, as evidenced by the fact that while car passengers often get motion sickness, car drivers almost never do.

On the Street Where You Live

Watching the other day as pedestrians passed me and my car by while I was stuck in city traffic, I wondered, how much faster does city traffic move now than in horse-and-buggy days?

Not much, alas. The average speed of a horse-drawn carriage traveling in a big city a century ago was about eight miles per hour. The average speed of a car driving through New York City today—accounting for traffic lights, pedestrians, double-parked trucks, and normal traffic congestion—is only 9.9 mph. In San Francisco, the most congested major American city, the average speed is even slower.

However bad that sounds, the worst major city in the world to travel through is Calcutta. This is according to the international courier company DHL, which makes it its business to know such things. To travel 5 kilometers (3.1 miles)

through the jostling crowds of humans and livestock, it would take you an average of four hours and thirty minutes—roughly two-thirds of a mile per hour.

If it's any consolation, traffic's been an urban problem for millennia. In the first century B.C., Julius Caesar banned chariot traffic from Rome during daylight hours.

When was the first traffic light installed?

Both Detroit and Cleveland claim to be the home of the first traffic light. We do know that both cities installed traffic lights in 1920, apparently independently of each other, and both inspired by the red, yellow, and green safety-light combo used by railroads.

However, London beats them both. In 1868, long before motorcars were even invented, the city government decided to bring some order to the chaotic procession of horse-drawn buggies, wagons, and pedestrians near the Houses of Parliament. It installed a revolving gas lantern with red and green signals on its sides, manually operated by a bobby who turned the globe with a lever at its base.

Unfortunately, the light proved about as dangerous as the traffic. On January 2, 1869, it exploded, injuring the police officer on duty and sending traffic-light technology back to the drawing board for another half a century.

Is there some mathematical formula they use to time how long before the Walk light changes to Don't Walk?

Most American cities use the recommendations of *The Traffic Engineer's Handbook,* which presumes an average speed of 6 feet per second for Joe Pedestrian. If you would like this expressed as a formula, take the total number of feet across the street and divide by 6. Subtract that figure from the total amount of time that the light will be green. That's how long your Walk signal should last. So if the street is 42 feet wide and your green light in that direction is 30 seconds long, the Don't Walk signal should start flashing at about the 23-second mark.

That is the case in most intersections, but not all of them. The *Handbook* recommends that in areas with a high number of senior citizens, traffic engineers should lower their pedestrian speed estimate to 4 feet per second.

See It Wiggle, Watch It Jiggle

Do the tidal waves that come after an earthquake swamp a lot of ships at sea?

It's true that earthquakes can create tsunami waves that often devastate distant coastlines. Strangely, though, they don't do much damage until they actually approach land. That's because in the open sea, a tsunami is a lateral wave that travels far beneath the surface of the water. Traveling underwater at great speed, it barely affects the surface of the water until it is forced upward by the shallows of a continental shelf. People obliviously dancing a fox-trot on a ship's deck would likely not even notice a tsunami passing underneath—the effect would be an imperceptible raising or lowering of the ship by a few feet, lasting a few minutes to an hour.

Generally, a ship on the ocean isn't a bad place to be in seismically unstable times. Unless it's tied to a dock that's lurching around because it's attached to land, a floating watercraft is pretty well insulated from an earthquake's damage.

Howdy, Sailor!

It doesn't make sense, Jeeves! How can sailboats sail into the wind?

You're right, it goes against all common sense. Yet it happens. That's why, when sailing, it becomes necessary to mentally throw out much of what you know about the wind.

Forget paper blowing down the street. Forget leaves blowing across the surface of a pond. Sailboats don't work that way. If they did, the only way to use one would be to turn the sail sideways and let the wind blow you in the same direction. Instead, a sail works more like an airplane wing. The result is that the wind actually sucks the boat forward instead of blowing it backward.

Here's how it works. The billowing shape of the sail makes the wind on the outside curve of the sail flow farther and faster than that on the inside curve. Because of that rapid movement, the air molecules get spaced farther apart, creating a suction that's twice as strong as that on the inside curve, sucking the boat in that direction. This suction by itself would try to pull the boat sideways. Luckily, the keel (that board sticking out of the bottom of the boat) keeps the boat moving forward, and the skipper keeps the boat from tipping over by leaning out from the other side of the boat.

No sailboat can sail directly into the wind, but it can come pretty darned close, sailing at an angle only 12 to 15 degrees off from the line of the wind. To go directly into the wind requires "tacking" (zigzagging back and forth).

If metal is heavier than water, how can a steel ship float on water?

The easiest way to understand the science involved is to float a metal bowl. While the metal itself is heavier than water and would normally sink like a stone, the shape spreads the weight over a large area. So, while the metal's weight causes the bowl to sink somewhat, the bowl's rim is high enough that the bowl will not be totally engulfed by the water. A ship is not unlike a large bowl. The surface area is much larger than you'd expect if you only see the top portion sitting above the water.

Life's Ups & Downs

What's the likelihood that an elevator cable will snap, sending me plunging to my doom?

Virtually nil. Elevators have multiple safety features, including brakes to stop them immediately if a cable breaks. This is done mechanically, using variations of an idea dreamed up by elevator engineer Elisha Graves Otis, founder of the elevator company that still bears his name. His safety brake was a large, bow-shaped spring that attached to the car and connected to the elevator cable. When taut, the cable kept the spring flexed. However, if the cable broke, the spring would immediately flatten out, jamming its ends into notched guard rails on either side of the elevator and bringing it to an immediate stop. Good show, Mr. Otis!

Demonstrated first in 1854, it's a system still in use today, making elevators just about the safest form of transportation there is, boasting only 1 fatality for every 100 million miles traveled. (Stairs, in comparison, are five times more dangerous.)

Tracks of My Charioteers

A friend claims railroad tracks are based on the width between the ruts made by Roman chariots. Is this true?

No, but we know where your friend got this idea.

You have to love the Internet: a piece of misinformation gets posted as a joke to a small mailing list in 1994 and takes on a life of its own, believed by millions to be true as it continues to richochet and reverberate through cyberspace nearly a decade later. Here's how it began. On February 9, 1994, Bill Innanen typed out a reasonable-sounding joke essay about the United States' and Britain's "exceedingly odd" railroad gauge—four feet, eight and a half inches between the tracks—and claimed that the measurement was based on the ruts left by ancient Roman war chariots. He sent it to a small mailing list composed mostly of military research and development engineers like himself. It was a group he expected would appreciate his wry point that military project specifications live long beyond their practical usefulness.

Well, they appreciated it way beyond his wildest expectations. The story spread and mutated, traveling at hyperspeed and inserting itself into books and news stories. Bill Innamen watched in bemusement, powerless to stop his snowballing creation.

So what's the truth? The truth is that there were dozens of track sizes used by railroads in both countries, but these eventually became standardized to the most popular gauge so that all trains could run on all tracks. But why that "exceedingly odd" gauge of four feet, eight and a half inches? It turns out that it isn't so odd after all. Gauge is measured from the inside of one track to the inside of the other; however, if you measured from the outside to the outside, you'd find that it equals five feet—certainly not an odd measurement at all.

> Where can I see the original train-and-chariot-tracks e-mail?

Zoological Zone

Step lively, there are wild animals in this chapter. From felines to canines, from monkeys to donkeys, we've got the answers to your questions about the wild kingdom.

Canine Conundrum

Why do dogs' hind legs move when they are scratched in certain spots?

For the same reason a person get shivers when someone touches his back or shudders when fingernails are scraped down a chalkboard. It's a reflex. Some experts say it's a method that dogs' wild ancestors once used to scare away predatory animals looking for a dog-sized lunch. Others say it exists to keep a dog from drowning–when nerves get stimulated on the sides of its body, a dog begins to "paddle." Whatever the original reason, this reflex has provided hours of entertainment for some pet owners, though it isn't much fun for the dog.

Sweat Blood and Tears

I've heard that hippos sweat their blood. How do they do this?

They don't actually sweat blood, but this erroneous belief is based on an oddity that bears mentioning. A hippopotamus has no oil or sweat glands on its entire body; it does, however, sport a gland that only kicks into action when it gets excited or nervous. This gland excretes a reddish, oozing liquid that is often mistaken for blood and coincidentally may function, in part, to scare off attackers.

Genetically, are the rhino and hippo in the same family?

You'd think so, wouldn't you? Both are large, ill-tempered, stumpy-legged herbivores. The hippo is second only to the elephant in the category of "heaviest land mammal"; the rhino is third.

However, they are not related. The rhino's habitat may vary, depending on the species, but they are exclusively land-dwellers. The hippo, in contrast, spends over half of its life in the water. Biologically, its closest living relative is the pig.

Primate Colors

Do apes and monkeys see the same colors as humans?

As far as scientists can tell, most primates see pretty much the same as people do. However, many of the New World monkeys are an exception to that–they don't see red well, leaving their world colored with hues of blue, green, and gold.

Are they called orangutans because of their color?

No. *Orangutan* means "person of the forest" in the Malay language.

What does **Homo sapiens** *mean?*

"Wise human." However, because anthropologists now identify other ancient subspecies of *Homo sapiens* (for example, *Homo sapiens neandertalensis*), modern humans are now known in the scientific world as *Homo sapiens sapiens*. This, of course, means "wise wise human," which seems to be overstating the matter.

Yipes, Stripes

Zebra stripes are supposedly camouflage, but where in Africa do you find black-and-white-striped terrain?

Would you believe that they can hide behind picket fences? We didn't think you would, but it was worth the try.

The stripes on a zebra make it quite camouflaged, even in its desert environment. The predators that prey on it—lions, hyenas, leopards, and cheetahs—don't see the same colors we do; the black and white creates a "disruptive discoloration." That means that the black stripes break up the outline of the animal, preventing many predators from being able to see the zebra clearly, if at all. This is especially true during the low-light times of day when many of their natural enemies hunt.

Head and Shoulders above the Rest

What other animals besides humans can stand on their heads?

The Asian elephant is the only other animal that, with training and a great deal of coaxing, has demonstrated this ability.

How many more neck bones are there in a giraffe than in a human?

The giraffe has highly elongated neck bones, giving it its great height, but a giraffe has no more and no fewer vertebrae than a human. Both a giraffe and a human also have the same number as a mouse, a pig, a dog, or an aardvark. All mammals except for two—the sea cow and certain sloths—have exactly seven vertebrae in the neck.

Size Counts

Is it true there used to be gigantic sloths in South America? When did they go extinct?

Indeed, a gigantic ground sloth called Mylodon was very much alive and living concurrently with humans during the Pleistocene epoch—or the Ice Age—some ten thousand years ago. It became extinct as recently as five thousand years ago. Except for the fact that the Mylodon had no tail (and was about the size of an elephant), it was very much like the modern-day sloth.

Which animal has the biggest penis?

A whale, though its penis is not easily measured to exact inches. Whales don't have erections exactly, but projections— the penis is usually hidden except during intercourse, when it is still partially hidden. But somehow those crafty marine biologists were able to measure it, and here's what they discovered: the right whale's penis measures over seven feet long, and its testes weigh about a ton.

Questions Worthy of Rumination

Do cows really have four stomachs?

No. A cow has one stomach, but it has four chambers. Those chambers are a necessity, because grass is very hard to fully digest. It takes a lot of nutrients to make a contented cow, so a typical cow takes in 100 pounds of grass and 300 pounds of water a day. The food first goes into the *rumen,* the first chamber of its stomach. After digesting for a while, the cow regurgitates the cud and chews it to break down the grass even further. After that, it swallows the mess into its second stomach chamber, the *reticulum.* From there, the food passes leisurely through two other chambers, the *masum* and the *abomasum,* before cruising down the intestines and back into the meadow.

Do cowslips have any connection to cows, or is it just a name similarity?

Cowslips are called that because they have become dependent on cow poop left behind in grazing meadows.

Sooiee!

Is there a biological difference between a pig and a hog?

It's all poundage. Any swine below 180 pounds is called a pig, and anything above that is called a hog. That's in the United States only, though. In England, all swine are considered pigs, whatever their weight. There are other specialized names for pigs during their various stages of life: a mother sow *farrows* (gives birth to) piglets; a *shoat* or *weener* is a child pig; a half-grown pig is a *barrow* (male) or *gilt* (female); adults are *boars* or *stags* (male) or *sows* (female). Castrated male pigs are also called *barrows*. What do you call a group of pigs? A *wallow.*

Beast of Burden

What's the difference between a donkey, a mule, a burro, and a jackass?

First there was the wild ass. A relative of the zebra, the African ass became a beast of human burdens thousands of years ago. This domesticated ass is what we came to call a donkey. It quickly spread to Europe and Asia. Burros, often used as pack animals, are just small, surefooted donkeys.

A male ass or donkey is called a jack (hence, "jackass"), and a female is called a jennet or a jenny. If a jack donkey is mated with a female horse, the offspring is a mule. A cross between a jenny and a male horse produces a hinny. Both mules and hinnies are hybrids, which means they can't reproduce.

Scrambled Eggs

Which came first, the chicken or the egg?

That's easy: the egg came first. Chickens have been traced back genetically to an earlier bird in Indochina called the "red jungle fowl." At some point one of these jungle foremothers laid an egg that had genetic mutations within it. The mutations were severe enough that the bird that hatched would have to be considered more of a chicken than a red jungle fowl. Hence, the egg came first.

How long ago was that? We're not sure, but we do know that people first domesticated the chicken about eight thousand years ago in Thailand.

How does an unborn baby chick breathe inside the egg?

An eggshell may look solid and impermeable, but it has about 8,000 pores that are large enough for oxygen to flow in and carbon dioxide to flow out. It was John Davy of Edinburgh, Scotland, who proved this in 1863 by pumping pressurized air into an underwater egg and watching thousands of tiny bubbles float to the surface.

Why are some chicken eggs brown?

Brown eggs are laid by rust-red chickens, most notably the Rhode Island red. White eggs are laid by white chickens, notably the white leghorn, which makes up about 90 percent of the U.S. egg-laying population.

A Very Odd Bird Indeed

Do flamingos come in different colors besides pink?

Yes. Flamingo babies, for example, are born white with gray streaks and take one or two years to develop their pinkish color. And depending on which of the five flamingo species you look at, the color will vary in intensity to the point that some of the lighter species look almost white. The best-known species in North America is the greater flamingo, which has a very deep pink hue. The pink comes directly from algae in the birds' diet that is heavy in carotene, a natural food color also found in carrots and other reddish vegetables. Their staples include small fish, insects, crustaceans, and certain types of algae. Some zoos give their flamingos carotene supplements to keep them "in the pink."

Why did the dodo become extinct?

The dodo, a relative of the pigeon, settled on an island named Mauritius millions of years ago. Because the island housed no predators, the ability to fly had no evolutionary benefit, and the bird eventually became flightless.

Dodos lived in relative peace for over 4 million years. In the 1500s, sailors began using the island just east of Madagascar as a stop on their trade route. Not long after, the Dutch set up the first human colony, bringing pigs, dogs, rats, and other animals. The animals loved the taste of dodos and their eggs. The flightless birds were easy to catch, and the entire species was wiped out by 1681.

But don't weep just for the poor dodo. Dozens of other bird species were exterminated when settlers cut down the Mauritius Island forests for sugarcane plantations in the 1800s. And don't forget the poor dodo tree, so named because of its interdependence with the bird. The birds ate its fruits, and the tree's seeds became dependent on passing through the bird's digestive tract before they could germinate.

Luckily, trees live longer than birds, so more than three centuries later, the dodo tree isn't extinct . . . yet. But its numbers have dwindled to a handful of ancients, and no new trees have germinated for more than three hundred years. However, there's reason to hope. Scientists have discovered that turkey gullets can work nearly as well as dodo gullets in germinating the seeds, so maybe the dodo tree won't go the way of its feathery namesake after all.

Because Q Was Too Hard

Why do migrating birds fly in a V formation?

Because it gives them the best of both worlds, reducing air resistance while allowing the geese or ducks in the back to see where they're going. Think of the V formation as the front of a boat cutting a path through water. The first fowl in the V formation cuts through the air and blocks some of the air and wind resistance for the two birds behind it, allowing them to glide through the air using less energy. Those birds do the same for the ones behind them, and so on all the way through the V. In this way, the birds can travel long distances with fewer rest stops during migration.

When the front bird gets tired, it drops back, and another takes its place at the front.

One Good Tern Deserves a Hover

What bird, animal, or fish migrates the farthest in a year?

The arctic tern, by a long way. It migrates 22,000 miles each year, flying from the Arctic Circle to Antarctica–and then, half a year later, flies back again.

Don't Bug Me, Man

How many insects are there for every person on Earth?

One estimate pegs the number at about 10,000 bugs for every human being, which comes as no surprise to anyone who goes outside on a hot summer night. Others say that estimate is way too low. Over 1.5 million known insect species populate the world today, but entomologists believe there may be millions more out there waiting to be discovered and classified.

How many mosquito bites would it take to completely drain your blood?

If you're an average adult, about 1,120,000.

Bee All That You Can Bee

Are killer bees more poisonous than regular bees?

"Killer bees," properly known as Africanized honeybees, are not more poisonous than other honeybees. They are, however, much more aggressive than the gentle European honeybee when they feel threatened or encroached upon, and are more likely to kill people or animals by bombarding them and delivering multiple bee stings.

There's a reason behind this aggressiveness. Conditions in Africa are rather harsh compared to the soft, moist climates of Europe and the Americas. To compensate for an environment that often has little food and water, these bees developed a higher level of aggression and the ability to produce a whole lot of honey on short notice.

That high-volume honey production was too good to pass up for apiculturists (beekeepers). In 1956 a geneticist named Warwick Kerr brought African honeybees to Brazil in an effort to crossbreed them with more docile European honeybees. His plan didn't quite work so well–his bees and their offspring never eased up on their aggressiveness. Worse, they escaped from captivity and started displacing the gentler honeybees. South American beekeepers simply got used to working with the more aggressive bees by taking greater precautions and using thicker bee suits.

My Honey's Eyes

What, exactly, do bees see? Are they blind?

Not at all. Here's a Web site designed by a neuroscientist that shows how and what scientists believe honeybees can see. Download the program and view the site's images, or submit your own to be changed into a bee's-eye view: http://cvs.anu.edu.au/andy/beye/beyehome.html.

How do bees communicate to each other where the flowers are?

They use a bee dance. It's a pattern they trace in a figure-eight shape, while wiggling periodically and flapping their wings. What tells the story is the speed of the bee, the direction the bee points, and the sounds the bee makes with its wings.

Here's how it works: Let's say a bee finds a flowering bush about fifty yards from the hive. After landing on a flower, she will bring some of the scent home on her legs. Inside the hive, she positions herself on the wall of the hive. Using "up" to stand for the position of the sun, she points her body in the direction of the bush. This shows her sisters where the bush can be found. The bush is only fifty yards away—pretty close for a bee—which means she must then wiggle frenetically in her figure eight to indicate that it is close. The other bees use these directional/distance indicators to find the dancing bee's bush. If they find it's worth dancing about too, they'll collect from it and come back and perform the same choreography.

Bug Lights

What are glowworms?

Glowworms are lightning bugs (or fireflies, if you prefer) during their larval stage.

What's in a lightning bug's lighter?

Lightning bugs produce light through chemical reactions in little cells called photocytes, using a substance called luciferin. When the enzyme luciferase is also present, the substances oxidize, creating a lot of energy. A spark of light is caused when the substances settle back down again.

It is still unclear why fireflies have this mechanism. It could be for mating purposes, to serve as a warning, or to attract prey. The lightning bug is capable of controlling the speed and length of the flashes, and although bug specialists haven't deciphered the code completely, the changes in pattern probably do change the meaning of the flashes.

Ladybug, Fly Away Home

How many ladybugs will stay in your yard after you release them?

Almost all of them will go. If you have enough aphids for the ladybugs to eat to make it worth their while, about 5 to 10 percent will stay. It helps to have water readily available near where they are released—some suggest watering the yard the night before the release. An average ladybug purchase should be about three thousand ladybugs for about a quarter of an acre of yard. And because of aphid breeding cycles, some experts say a second spreading of ladybugs may be necessary not long after the first.

Super Fly

Didn't doctors once use maggots to cure infections?

Doctors prescribed maggots to help heal wounds during World War I. How did this happen? The doctors caring for war casualties noticed that some patients' wounds healed quicker and were more resistant to infection than others. At closer look, they realized those patients had flies landing on their open sores . . . and laying eggs. Before long, "flyblown" wound treatment became all the rage. Flies and their larvae seem to contain some healing and antibiotic properties that medical professionals couldn't ignore, especially not under wartime circumstances. Still, to make the conditions a bit more safe, doctors began raising maggots and sterilizing them before placing them on open sores.

Thankfully, science soon revealed the healing substances–allantoin and urea–and learned how to extract them, bypassing that menacing go-between. Despite their laudable war effort, flies are also known to be carriers of typhoid, cholera, salmonella, dysentery, leprosy, tuberculosis, and many other life-threatening diseases.

Check Out My Web Site, Little Fly

Why don't spiders stick to their webs?

Some spiders do. A spider's web is constructed of two types of silk: "anchor" and "snare" threads. The anchor threads are used to construct the basic web; the snare threads are sticky and are used to trap insects. To get around the web, the spider uses the nonsticky anchor threads. Once in a while, though, a spider will inadvertently catch a leg or two in a snare thread. When this happens, she simply secretes an oily solvent to free herself. Some spiders have a special claw called a scopula, at the bottom of the hind legs, that they use to slide along the sticky snare strands and help them escape from their own homespun death traps.

Do spiders suck insects' blood?

Yes. But they don't just suck out an insect's blood, they usually suck out the guts as well. Most spiders inject into their prey a venom that contains toxins that paralyze. Then they pump digestive juices into the victim and either suck out all the victim's juices right then and there, or wrap up the live-but-paralyzed casualty to keep it fresh until the spider's ready to eat it at a later time.

Kiss of the Spider Woman

Can you die from a black widow spider bite?

You can, but most people who have been bitten don't. Although a female black widow is not particularly aggressive (and the males are harmless), she will bite in self-defense and is quite poisonous. The effect of her bite is painful and frightening, and there isn't much that can medically be done to reverse the process.

What are the symptoms of a black widow spider bite? You start with a stinging pinprick at the location. Dull numbing pain soon follows, often accompanied by some swelling. Within about thirty minutes, severe stomach pains and clenching of the abdominal muscles begin. After a while, you may get spasms and severe pain in the arms, legs, and feet. Finally, the most frightening part: paralysis or partial paralysis, chest constriction, and difficulty swallowing.

Although there isn't much that can be done about it once it's happened, seeking out a doctor is a good idea anyway. The symptoms can be alleviated somewhat by medical professionals. The good news, unless you're one in a hundred, is that the black widow spider bite has only about a 1 percent fatality rate. Those most at risk of dying are the very young, the old, the ill, and those with allergies.

Can spiders lay eggs in skin?

No. Other insects may do this—like the mites that create the skin disorder scabies—but spiders are far too concerned with the well-being of their offspring to entrust them to the likes of us. Warmth isn't what mother spiders are necessarily seeking for their egg sacs, anyway—security is.

Multiply by Dividing?

Is it true you can cut an earthworm in half, and both halves will grow back?

If it were true, then worms could dispense with that messy sex business. Worm sex *is* messy, although pretty efficient. Worms are hermaphrodites, so they can mate with any other worm of their species (although *not,* as some would have you believe, with themselves). They do that in a slimy sodden mess, by writhing, putting together their clitella (those enlarged bands near worms' heads), and exchanging sperm. Afterward both partners will likely be pregnant and each can lay an egg capsule about a week later. After fourteen to twenty-one days, one to five baby worms will be found under the pumpkin patch.

The reality, of course, is that the "multiply by dividing" theory is a myth. While some worms can regrow a tail if they lose it, the part that's lost will die. True, the lopped-off tail can wriggle around helplessly for a few hours, but that's just dead worm walkin'—somewhat akin to a chicken running around with no head—and the tail will eventually stop writhing.

The other end might also die from the injury, but at least it has a chance of surviving if its intestines and other vital organs are still intact.

Is there a name for a fear of worms? Helminthophobia.

Cold Comfort

How do bugs survive in Arctic temperatures?

It's a good question. It gets unbelievably cold in the Arctic—down to an average temperature of 90 degrees below zero, with a windchill factor like you wouldn't believe. The animals living in the Arctic have developed strategies to help them beat the cold. These include ears and tails that reduce heat loss, lots of body fat, small heads, thick fur, and hibernation in the winter. The bugs use some of the same strategies, plus some unique ones that the birds and mammals haven't figured out yet.

For example, take the Arctic woolly bear caterpillar. It's unusual because not only does it survive the long, cold winters, it does so for fourteen years before reaching adulthood. It's the longest-lived caterpillar on Earth. Above its tiny pink legs, the woolly bear is fuzzy to preserve heat, and dark to absorb solar heat. That's not so strange. But what is strange is that it doesn't avoid freezing, but instead controls the process by manufacturing glycol within its body.

What normally happens when fluids freeze inside cells is that the sharp ice crystals rupture the cell's membranes, killing the cell. Glycol, a type of alcohol that is sometimes used as antifreeze by humans, minimizes the size of the crystals and slows down the rate of freezing. The woolly bear's body freezes in a systematic pattern: first the gut, then the blood, and then everything else. However, the glycol keeps the cytoplasm inside each cell from actually freezing solid. (Antarctic spiders and beetles also make their own glycol to keep from getting freezer burn.)

Throughout its life, the woolly bear freezes and thaws a total of thirteen times. The hungry little caterpillar is only lively and active for the three weeks of the year right before the summer solstice, soaking up the sun's rays on the frozen tundra, grooming its frizzy hair, eating, and growing a little bit more before going back into its icy tundra bed again. And for what? To eventually turn into an *Arctia caja* moth for less than a year. It lives for only part of the Arctic summer, mates, lays eggs, and dies. So it goes.

In Black & White

What bird dives deepest underwater?

The penguin. Some species can dive 900 feet, the height of a typical seventy-five-story building, and hold their breath for nearly twenty minutes while swimming.

What if you got a cargo ship full of ice and went to Antarctica, filled the ship full of penguins, and transported several thousand to the Arctic? Would they survive?

Scientists say probably not. For one thing, they would find none of the landmarks that they used to establish their ancestral breeding grounds. For another, they don't defend themselves and their young well—they thrive so well in the Antarctic precisely because there are no land-based predators to disrupt their nesting. If you moved a colony of penguins north, the odds are that for a year or two they would provide cheap protein for polar bears and Arctic wolves, before their numbers steadily dwindled to nothing.

Could penguins ever fly?

We know it's hard to imagine flocks of penguins flying overhead, but millions of years ago penguins did fly. Eventually, though, flying became superfluous to their lives. They didn't have any land-bound natural enemies that would make fast escape by air necessary, and they spent most of their time in the water. Air wings gave way to flipperlike water wings, a change that made penguins able to "fly" through water using the same motions that other birds use to transport themselves through the sky. Swimming underwater, they can average about eight miles per hour (with bursts of up to 25 mph), "dolphining" up into the air every minute or so to breathe.

Nice Weather for Boys

How is sex determined in alligators?

Alligators, like many other reptiles, don't have sex chromosomes. As a result, alligator gender isn't determined by the interaction of the parents' genes. Instead, it is completely decided by temperature.

When it's egg-laying time, the female finds a nice large area on the bank of her swamp and digs a large hole, up to six feet wide. She then deposits up to fifty eggs and covers them with leaf debris and mud. The soil and composting leaves provide the warmth the eggs need to mature. During the two-month incubation period, the mother guards the nest from the water nearby.

Gender is determined during the second and third week of incubation. If the temperature of the eggs is 86 °F or below, the alligators inside will all become females. If the temperature is 93 °F or above, the offspring will all be male. If the eggs simmer at a more moderate 88 °F, as most nests do, the babies will be both male and female.

At the end of incubation, the first mature babies begin to make a barking noise from inside their shells. Barking alligators can be heard up to fifteen yards away, and this signals to Mom that it's hatching time. The mother's presence is crucial, as she uncovers the eggs from their buried home. If the mother isn't available or fails to hear the barking, the offspring can't hatch.

No matter what gender emerges, mother alligators continue to be very protective. For the first two years of life, babies travel with Mom, sometimes on her back or in her mouth, and she aggressively guards them from predators like otters, turtles, skunks, raccoons, and other alligators.

Hang in There, Baby!

Do the suction cups on lizards' feet ever lose their suction, sending the lizards plummeting to the ground?

The lizards that climb walls and ceilings don't usually fall off. But it's not because they have great suction. Actually, they're not really using suction at all.

Here's how their feet work: On the bottom, there are dozens and dozens of grooves. (You can see them if you look closely at the bottom of a lizard in a glass-walled terrarium.) On each of those little grooves that you can see, there are dozens more microscopic ones you can't. And on each of those microscopic grooves are thousands of hairlike bumps.

So what good do bumps do on a smooth surface like glass? Well, if you look at the surface of glass through a microscope, you'll see that it isn't so smooth after all: its surface is covered with microscopic pits, bumps, and grooves. To a lizard, tile, concrete, Sheetrock, and even glass all look like nice big rock-climbing walls with plenty to hang on to. A lizard's pads will grab onto the surface and hang on tightly, even if the lizard is scared, tired, sick, or dying. In some cases even death isn't enough to dislodge a lizard from a wall. Sometimes it takes hours for a dead lizard to drop because the feet are still clinging onto the tiny imperfections.

Bad Newts for Predators

Is a newt a type of salamander, or is it in a different category entirely?

Yes, a newt is a type of salamander in that it fulfills the basic requirements: its skin is thin, it breathes through gills until its lungs grow, and the adults can leave the water and walk among us as land-dwellers. The biggest difference from other salamanders is that adult newts have flatter tails than most. This isn't a big difference, we know, but it will have to do.

When a newt finally grows up and leaves the water with a working set of lungs equipped for a life on land, it's called an eft. From then on, they return to the water only briefly to catch up with old friends, check the mail, and mate.

Adult newts are brightly colored and are poisonous to predators. We can only guess that this is why "eye of newt" is part of the witches' recipe in act 4 of Shakespeare's *Macbeth* ("Double, double toil and trouble; Fire burn and cauldron bubble").

Sometimes It's Better Not to See

How come frogs blink when they swallow?

Frogs don't have teeth, so blinking helps them get their food down. Frogs' eyes, not unlike those of the hippopotamus, are big, bulgy, and located on the tops of their heads. This helps them to see, even when their bodies are submerged in water.

The problem is that it also means these huge eyeballs take up a lot of space inside the frog's head. When the frog closes its eyes to swallow, the lids push them into the back of the frog's mouth, helping to force the wiggling bug down its throat. Since their tongues are better designed to shoot out and retrieve food than to push food into their stomachs, this system isn't a bad one. And frankly, if we had to eat bugs, we'd probably close our eyes, too.

Vitamin A Delivery C.O.D.

Cod-liver oil showed up a lot in old movies and books as something parents often gave to their unwilling kids. How come it isn't used anymore?

Both shark oil and cod-liver oil were once very commonly used as supplements for vitamins A and D. However, in the 1940s scientists figured out how to cheaply manufacture the vitamins, saving generations of kids from spoonfuls of the fishy, slimy swill.

Fish in Chips

My uncle caught a fish, and it exploded like a balloon. What was that about?

Here's why it happens: Most fish keep their equilibrium while in the water with an air bladder that balances out their body mass and makes them essentially weightless, neither sinking nor floating to the surface. This is energy-efficient for them, so they don't have to continuously flap their fins and tails to stay at a desired depth.

A deep-swimming fish needs a lot of gas in its bladder to withstand the extra water pressure of deep water. While fish can make subtle adjustments to the amount of air inside their bladders, they can't do it quickly enough if they're caught on a hook and dragged suddenly toward the surface. As their bodies rapidly pass through the water from a lot of pressure to a middling amount, their bladders expand. This is sometimes enough to kill them before they even reach the surface. Finally, when the fish is pulled out of the water into the low pressure of the air, blam! Fish pieces everywhere.

Not all deep-sea creatures explode, however, because some don't have bladders. You can tell because they drift to the bottom when they stop moving. Sharks don't have them, and neither do skates or stingrays.

This Racer Leaves Squid Marks

If I were going to a cephalopod race, should I bet my money on a squid or an octopus?

Squid. Octopi are smooth, gliding along on their tentacles like an eight-legged Fred Astaire, but they're no match in speed for a squid. Squids can shoot water backward like a rocket engine, pushing them forward in great bursts of speed. Some can glide through the water at about 35 mph, and can even leap right out of the water.

Does an octopus's ink serve any purpose besides darkening the water for camouflage?

Standard procedure is for the octopus to squirt the ink, change its own color as camouflage, and simultaneously jet away under the inky murk's cover. While the ink does do a good job of this, it also holds a secret weapon that further facilitates an octopus's escape. The ink, even when diluted in the ocean, is poisonous to some creatures and has a stunning effect on others. How poisonous is the ink? If an octopus in an aquarium released its ink, it would kill everything inside the glass walls—including the octopus.

An Eel That Gives a Lot of Juice

How much electricity can you get from an electric eel?

Oh, about 350 to 650 watts, but you better get a surge protector: an electric eel delivers only three to five bursts of electricity when it discharges, each lasting about one-five-hundredth of a second. While its voltage is enough to temporarily stupefy a human being, the eel normally saves its energy for the frogs and small fish that it eats.

By the way, the electric eel isn't really an eel at all, but a fish related to the carp and catfish. It is also not that unique—scientists have identified about five hundred other species of fish that can also generate electricity.

The Old Shell Game

Why do lobsters turn red when they're cooked? Is that the color of their blood?

It's not their blood. Lobster blood is colorless unless exposed to oxygen, at which point it develops a bluish tint. The shells of lobsters are mostly gray, green, or brown when they're alive, but never red. Boiling them breaks down the various pigments that color their shells. Their most durable tints are the reds, so they're the last to disappear during boiling.

Sea Monkey, Do Monkey

What water creature can live the longest time out of water?

The record is about 10,000 years, set by some brine shrimp. Also known as "sea monkeys" in toy stores, the little crustaceans go into a form of suspended animation when they get caught outside the salt water where they make their home. Some desiccated ones, determined to be 10,000 years old, were found by oil drillers near the Great Salt Lake in Utah. Although they demonstrated no measurable sign of life, the little guys came back to life when placed in salt water.

Medicine

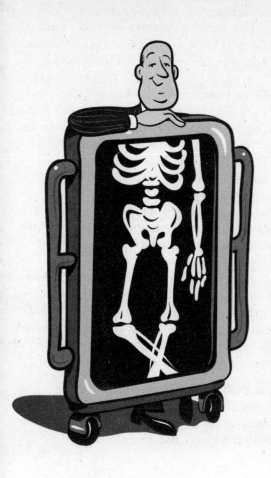

Healing powers were once considered so mystical that shamans were treated with awe and terror. Today, those who work in healing are still considered pretty mystical. After researching your questions about illness and medicine, we can understand why.

Thicker than Water

When was the first successful blood transfusion?

Before people discovered the mysteries of blood type, transfusions were a hit-or-miss procedure. Some historians believe that the Incas may have been the first to do successful blood transfusions. If that's true, what would have made the job easier is that they were all the same blood type (O positive).

The year of Columbus, 1492, witnessed the first contemporary-documented attempt at a blood transfusion. The doctor was trying to save a dying pope, Innocent VIII. Not only the pope died, but also the three boys from whom the doctor extracted blood. The doctor quickly went on the lam, and blood transfusions weren't tried again for nearly two centuries.

In 1665 Dr. Richard Lower of England successfully transfused blood back and forth between two dogs using feather quills. Unfortunately, he then tried a lamb-to-man transfusion. The man died, and research into blood transfusions was abandoned for another 150 years.

In 1818 Dr. James Blondell saved a man's life with the first documented person-to-person blood transfusion. Unfortunately, because people still had not yet unraveled the mystery of blood typing, his next attempts were failures. His patients died, and research ceased for yet another century.

Why is blood-typing important? Mixing incompatible blood quickly creates a sea of red dumplings swimming in yellowish liquid, bringing on instantaneous death. In 1900 Dr. Karl Landsteiner discovered the A, B, and O blood groups and demonstrated that blood could be successfully mixed within the same group. He named the differences the "Rh factor."

What does the Rh stand for in "Rh factor"?

It's short for "rhesus," in honor of the research monkeys that gave their blood (and sometimes their lives) as Dr. Karl Landsteiner tested out these properties.

Pump Up the Volume

When were the first artificial heart parts put into humans?

The first artificial heart valve was implanted in 1952 and used a tiny plastic ball to alternately open and close the blood flow. One problem was that when patients opened their mouths, clicks from the valve were annoyingly loud.

The first artificial heart was made of Dacron and was in a Texas man's chest in 1969. He died four days later.

Hobson's Choice

What's worse: a stroke, a heart attack, or a brain aneurysm?

You don't want any of them, but if you had to choose which *not* to get, pick the heart attack. Here are the reasons:

1. An estimated 40 percent of all heart attacks are fatal. A heart attack occurs when blood flow is cut off to and from the heart. Over 220,000 people a year die from sudden cardiac arrest, making it the leading cause of death in the United States.

   As mentioned on page 40, if you watch those medical TV shows, you may get the impression that when someone is given cardiopulmonary resuscitation, or CPR–doctors pounding on patients' chests to get their hearts restarted–they have a pretty good chance of being revived. This isn't so, say the experts–television depicts about a 67 percent survival rate among people whose hearts have stopped; in real life, it's less than 15 percent.

2. Stroke is the third leading cause of death in the United States, with just under 200,000 people dying from it annually. With a stroke, vital blood flow to parts of the brain is stopped. This happens either when a blood clot blocks off part of a blood vessel or when a blood vessel actually bursts. The survival rate of stroke is much higher than for heart attack. About 85 percent of all stroke victims survive. However, only 10 percent of survivors completely recover. The others are left at least partially impaired.

3. An aneurysm is a place in a weakened arterial wall that is bulging out (think of a bike inner tube with a weak spot, and you get the idea of what it looks like). An aneurysm can occur in several locations within the body, but most occur in the brain (cerebral aneurysm) or the abdomen (abdominal aortic). The problem with these bulging weak

spots is that they sometimes burst. In the brain, that can be catastrophic. It's estimated that 1 to 5 percent of all Americans may be living with brain aneurysms. The good news is that there's a 1 in 100 chance that an aneurysm will explode. However, if that happens, there's only a 50 percent chance of survival and a 25 percent chance of escaping severe brain damage.

Open up Your Heart & Let the Sunshine In

When was the first open-heart surgery performed?

One day in 1893, a Chicago man named James Cornish was stabbed in the chest and rushed to Provident Hospital, bleeding to death. Despite the medical profession's lack of success in fixing damaged hearts, Dr. Daniel Hale Williams rushed Cornish into the operating room and called in a team of surgeons.

At the time, attempts at heart surgery invariably ended with the death of the patient for a variety of reasons. Injuries to the heart are mortally serious and don't allow the surgeon much time to work; even without time pressure, doing precision cutting and stitching on a still-beating heart is a near impossibility. Still, Williams began the painstaking task of entering Cornish's chest and repairing the knife damage to the right ventricle. Somehow the team managed to successfully suture the rip and close Cornish's chest without him dying from either the injury or the surgery.

Miraculously, despite the unavailability of antibiotics at the time, Cornish lived and left the hospital a month later, making Dr. Williams the first person to successfully perform an open-heart surgery. President Grover Cleveland heard about the incident and appointed Williams as head surgeon of Freedmen's Hospital in Washington, D.C.

Catgut Your Tongue?

How much surgical catgut can you get from one cat?

None. Despite its name, catgut isn't actually made from the guts of cats. It was originally derived from the intestines of sheep, which were dried and twisted into thin, strong strands. While catgut could be made from the intestines of cows and other animals, there's no evidence that cats were ever its source.

So why isn't it called "sheepgut"? There are several theories, but the one we like best is that the misnomer came into use at a time when violins were strung with the intestinal strings. "Kit" was another name for a fiddle.

Regardless, catgut is on its way out as a surgical suture. Other biological sources for sutures include silk and hair. Since the outbreak of bovine spongiform encephalopathy (BSE, or mad cow disease), doctors and patients have worried about catgut carrying the disease. Synthetic materials have proven superior to biological catgut over the last decades; the worry over contracting BSE from natural sources has further hastened this trend.

Gathering No Moss

Besides the location, is there a difference between kidney stones and gallstones?

There is. The "stones" that grow in each are related to what each organ does for the body.

The gallbladder is a small sac that lies just below the liver. It holds bile that's been produced by the liver, saving it until the stomach needs it to add a dollop of bile during digestion. Sometimes the stuff that makes up bile–cholesterol, recycled red blood cell pigments, traces of calcium–clumps together and hardens into a solid.

Cholesterol is the leading ingredient in gallstones. When your body contains too much cholesterol, the excess can begin to form solid particles in places where you don't want it. In the gallbladder, these fatty, waxy clumps can be tiny and cause no problems at all. However, they can also get as big as golf balls, completely blocking the duct that leads from the gallbladder to the stomach, requiring immediate medical attention.

Kidney stones are formed from the solids found in urine—primarily uric acid and calcium compounds. They can get up to the same size as gallstones, but tend to be smaller. When kidney stones get large and solid enough to block the ureters (the tubes that lead down and into the bladder), medical attention is needed immediately so that waste products don't back up in the kidneys.

What if you want to get your stones polished up and mounted on jewelry for a novelty (see below)? Note that gallstones are usually brownish green in color and smooth, while kidney stones are usually brown or yellow (or a combination of the two); they can be jagged, sharp, or smooth, depending on how they formed. Some kidney stones are quite striking in appearance, actually—a nice mix of colors and interesting shapes.

Where can I see pictures of what kidney stones look like?

What's kidney stone jewelry? Is it really made from kidney stones?

No, but it can be pretty confusing. What's called a kidney stone in the jewelry business is really nephrite or hematite. The stones have been historically used by crystal healers and others in "treating" problems with the kidney, hence the name kidney stone.

That doesn't mean, though, that gall- and kidney stones haven't been used to make jewelry. For example, take Larry Hagman's ring. When television actor Hagman (*I Dream of Jeannie, Dallas*) had a liver transplant, he also decided to have gallstones removed. Although most hospitals refuse to allow patients to take home their spare parts, celebrity has its perks. Hagman reclaimed his gallstones and had an artist friend turn one of them into a ring. Hagman doesn't wear the ring, though—he says it's too soft and will break. Instead, he says, he keeps his once-internal keepsake inside a velvet box, hidden away in a "special place." No doubt his heirs will thank him when they inherit it.

Gastro-Butterflyus

Is there a medical diagnosis for having a pit in your stomach?

Not a diagnosis per se, but there is a name for the condition. It's called an epigastric sensation.

A Happy Glow

How much radiation do you get when you're x-rayed?

A regular, routine chest X ray gives you the equivalent of three days' worth of background radiation—the levels of radiation you receive naturally from the sun, Earth, and other celestial bodies over the course of three days. A full-body CAT scan gives you the equivalent of four years of background radiation. This means that unless you're getting a full body scan regularly, the benefits of X rays probably vastly outweigh their risks.

Hey Mickey, Lend Me an Ear

What's the weirdest genetic crossing that medical researchers have performed in labs?

There's so much genetic work going on, some of which is still secret, that it's hard to really determine which one is the weirdest. As far as we're concerned, one of them has to be the glow-in-the-dark mice that obtained this ability courtesy of glowing jellyfish DNA transplants. Another is the pig teeth that grew inside a rat's intestines. Then there are the spider-silk genes implanted in goats' mammary glands, allowing the goats to produce surgical-grade spiderwebs along with milk.

However, when it comes to finding the most disturbing-looking example of cross-species cross-dressing, there's no competition at this writing: it's the experiment of a mouse with a human ear growing on its back.

To make it happen, doctors constructed what they called a "scaffolding"—a structure that looks like a human ear—using a new biodegradable plastic. They seeded it with human cartilage cells and placed it under the skin on the back of the mouse. The blood supply of the mouse gave the cells the warmth and nourishment they needed to thrive and grow around the ear-shaped scaffolding. The purpose of the test research was to see if doctors could grow ears and noses for patients who were born without one or the other, or who lost one in an accident.

If things work the way they're supposed to, the ear cells will grow around and then replace the scaffolding, and the fully formed ear can then be removed and transplanted onto a patient. Unlike many such procedures, the mouse can theoretically have a life following the procedure; its death is not required to harvest the ear.

You can see a picture of the famed ear/mouse at http://tlc.discovery.com/convergence/superhuman/photo/zoom_03.html.

Getting into Your Genes

Is a genome the same as a gene?

No, a genome is the DNA—or all of the genes within an organism.

What is the Human Genome Project?

Originally proposed by the Department of Energy in 1990, the Human Genome Project set out on a fifteen-year biological journey, to discover all of the human genes. By early in 2001, both a government group and a private group of scientists had done an initial rough draft of the human genome. As of this writing, the scientists expect that in 2003 they'll be able to publish a final draft of the genome, pushing the targeted completion date up two years. Identifying all of the 30,000 or so genes and the 3 billion chemical pairs that occur within human DNA is no small feat.

But that still won't be a time for champagne and backslapping. Most scientists say that only when the genome is completely mapped will the real work begin. Researchers have yet to learn how they can fully utilize the knowledge to discover what all those genes do and how that information can help humanity (for example, to eradicate some of our hereditary diseases). Tools to make specific changes within isolated genes have yet to be created, and ethical standards for the work are still being discussed and debated.

At this writing, only the genes in chromosomes 13, 20, 22, and Y have been thoroughly investigated and mapped. By the time you hold this book in your hands, many more will be mapped, and perhaps the final draft will be completed. At the rate in which genetics is growing, by the time your grandchildren hold this book in their hands, perhaps gene-altering drugs will exist as over-the-counter items, and the Human Genome Project will only be a mention in the medical history books.

Not Losing Your Temperance

How does the anti-alcohol pill work?

A couple of pills are specifically used in the treatment of alcoholism. The latest one is naltrexone HCl (brand name ReVia)—a pill that blocks the effects of opoids. Opoids are neurotransmitters that the brain releases when thrilled—and also when drinking. They bind to receptors in the brain, and are believed to cause the "high" that alcoholics attempt to achieve when they drink. The idea is that when opoids are blocked from binding to pleasure receptors in the brain, drinkers don't get the results they're looking for, so theoretically they stop drinking.

An older drug treatment for alcoholism is tetraethylthiuram disulfide, sold as the drug Antabuse. During the 1930s, workers in rubber-manufacturing plants discovered that they got violently ill whenever they consumed even small quantities of alcohol. Not long after, scientists discovered that the nausea came about because of tetraethylthiuram disulfide, a chemical used in rubber manufacturing. Apparently the drug changes the way the body metabolizes alcohol. It increases the amount of acetaldehyde, producing symptoms similar to that of a hangover: flushing, throbbing head, nausea, and vomiting. Since it's nonlethal and works even when only a small amount of alcohol is consumed, doctors began using the substance in the treatment of alcoholism.

It's The Little Things That Count

Where did cells get their name?

After Anton van Leeuwenhoek invented his microscope in 1642, a colleague was using it to look at thinly sliced cork. He thought that the neat rows of squares looked like the living quarters of monks—"cells"—and so the name stuck.

Since they're so tiny, when did people discover that sperm led to conception?

Of course, while people had figured out long before that sex produced babies, they hadn't yet discovered spermatozoa. In fact, it wasn't until the invention of the microscope that people even knew the little fellas existed. Anton van Leeuwenhoek, inventor of the microscope, became one of the first to see sperm, bacteria, and protozoa. However, he didn't make the connection between the wiggly little spermatozoa and conception. Neither did priest-biologist Lazzaro Spallanzani, who in 1779 filtered semen from amphibians to determine what exactly caused conception.

Spallanzani almost got it right. He discovered that the semen fluid became less and less effective at impregnation the more completely he filtered it. He also discovered that the residue filtered out—solids and spermatozoa—retained the power to impregnate. However, he came to the wrong conclusion: he decided that it was the mucousy solids that caused impregnation, and that the spermatozoa were merely some sort of parasite.

It wasn't until 1840 that Martin Barry opined that sperm actually entered eggs, causing conception. Finally in 1879, Hermann Fol, a scientist with a good microscope and plenty of patience, actually witnessed this happening.

In Hot Water Again

Does boiling water kill all germs?

Boiling water will kill most bacteria and viruses, including those that want to do you harm. However, if you're looking for complete sterility, boiling temperatures won't kill every type of microorganism. For example, heat-loving bacteria have been found swimming in superheated volcanic vents bubbling up from the ocean's floor, in the hot springs of Yellowstone Park, and even in geothermal power plants, surviving temperatures well above the boiling point of water.

Alternatives to Science

Is there a difference between homeopathy and holistic medicine?

There's a big difference. Homeopathy comes from the belief that bringing on a small amount of the symptoms that ail you will make you well–*homeo* means "similar" in Greek; *pathos* means "suffering." Founded in the 1800s by a doctor named Samuel Hahnemann, the practice was born out of Hahnemann's experiments with cinchona bark. The bark cures malaria because it's loaded with malaria-curing quinine but also mimics the disease in the process, making the patient run a high fever. Hahnemann decided that since the bark gave the patient a few malarialike symptoms in the process of healing, this must be the reason the patient was healed. Taking the logical next step, Hahnemann decided that most illnesses could be cured using the same model, so he set about looking for serums that would irritate the nose to cure hay fever, make the head hurt a little to cure headaches, etc.

This is admittedly hard for our hyperrational Western minds to accept, but homeopathic practitioners believe that the more diluted the tincture, the more effective it can be. In analysis of some homeopathic remedies, however, scientists have found little to no traces of the "curing" agent at all, leading to the speculation that most aches and pains, given time, resolve themselves no matter what you do.

Holistic medicine, in contrast, is based on the belief that you must treat the whole person, not just the problem. Holistic practitioners may prescribe activities like yoga, acupuncture, exercise, meditation, and massage as part of a person's treatment. This approach is becoming more accepted by mainstream medicine as science recognizes the negative role that stress, mental states, attitudes, and unhealthy lifestyle choices can play in illness. On the extreme edges of holistic medicine, however, are those practitioners who eliminate much of traditional medicine from their practices, refusing surgery and other proven medical interventions and helping discredit an otherwise intelligent approach to health.

But You Can't Take the Cancer out of Salem

If you took the nicotine out of tobacco, would cigarettes be safe to smoke?

Any amount of soot or smoke in your lungs and bloodstream can damage your health, regardless of the source or composition.

It's true, though, that some sources of smoke are worse than others. With about four thousand chemicals found in cigarette smoke, merely taking out the nicotine wouldn't be enough to make a healthy cigarette. For example, there's carbon monoxide–the deadly suicide gas produced by the tailpipe of your car. There's also carbon dioxide–the stuff your lungs expel when you exhale. There's even methane, the ingredient contained in flatulence.

Other potentially harmful ingredients include formaldehyde, the stuff used to preserve dead tissue in embalming; acetone, a common ingredient in fingernail polish remover; propane, a fuel used in grills and lanterns; hydrazine, a fuel used in rockets; toluene, a substance used in paint thinner, lacquer, and industrial adhesives; and hydrogen cyanide, often used in rat poison.

Garbage In, Garbage Out

If I smoke, how much more likely am I to die from lung cancer than someone who doesn't smoke?

If you smoke, you're anywhere from twenty to thirty times more likely to die of lung cancer than a nonsmoker. And you are much more likely to die of heart disease or another lung disorder than a nonsmoker, too. Nicotine patch, anyone?

Bad Medicine

What's the deadliest toxin that occurs naturally?

The bacteria *Clostridium botulnum*. It's the cause of botulism, the deadly food poisoning that is sometimes contracted from canned goods or meats.

Strangely, it's also a medication used in low doses for people who have involuntary muscle spasms. In very small quantities, botulinal toxin causes local paralysis, easing painful and debilitating muscle contractions. And of course it is this same low-dose medicine that is used in plastic surgery under the name Botox. Injected into facial muscles, it paralyzes them, reducing the appearance of fine lines and wrinkles. Some say it also reduces the appearance of emotion or personality.

> **How many people in the United States die each year from food poisoning?**
> Approximately eight thousand.

Do anabolic steroids really make you a stronger, better athlete?

Despite anecdotal stories of athletes healing faster and pushing harder as a result, science has been unable to definitely prove that using anabolic steroids actually increases performance. As a matter of fact, because steroids are psychologically addictive drugs, most doctors believe that at least some of the positive effects are in the users' minds.

Anabolic steroids resemble what's traditionally known as "male" hormones (although both sexes have them), primarily testosterone. Anabolic steroids were originally created to treat a condition called hypogonadism, which occurs when the testes don't produce enough testosterone on their own, leading to delayed puberty. Today, anabolic steroids are also medically used to treat some forms of impotence and wasting syndrome, caused by disease.

However, when they are used to improve athletic performance—usually through illegal means—many athletes discover that steroids have a dark side. In men, side effects can include infertility, breast development, and withered testicles. In women, side effects include hirsutism—increased hair on body and face—lowered voice, and baldness. Some of the more dangerous side effects for both sexes include weakened, rupture-prone tendons, liver cancer, rage, delusions, abnormal heart enlargement (particularly of the left ventricle), and heart attacks. Furthermore, illegal steroid abuse has often been linked to reused contraband needles that can, of course, spread infections such as hepatitis C, HIV, and AIDS.

So how come those in the medical profession don't know for sure if anabolic steroids can truly enhance athletic performance? Primarily because the dosages athletes are using illegally are simply too high to be considered safe. Doctors cannot ethically give participants—willing though they may be—the same dosages, and therefore cannot properly track the results.

See You Later, Incubator

I saw an old map of Coney Island. It had a spot marked "Incubator Babies." What's that about?

We're glad you asked, because the history behind the incubator is a little bizarre.

When Alexandre Lion invented the baby incubator in 1891, it was in response to an alarming infant mortality rate in France. Lion built a device with a water boiler and a fan system that blew warm, filtered air into a covered baby bed. This counteracted the problems premature babies have with maintaining their body temperatures. Positive results were seen instantly, and the incubator saved lives that normally would've been lost.

Lion appealed to hospitals for the support to build more, but they wouldn't come up with either the funds or the interest. He resorted to showing the incubators at exhibitions, but found that without the main ingredient—the baby—he could generate little interest in what essentially looked like an empty box. So he made a bold move. He solicited premature babies from local hospitals. The hospitals, believing preemies were going to die anyway, lent them to him, giving the babies a chance at life.

Lion's first live exhibit had a futuristic appearance, with wet nurses, incubators, and live babies behind a glass wall, allowing fairgoers the ability to walk by and gawk. They were so amazed, in fact, that many had to be turned away. Lion decided he would start charging admission to reduce the crowd size. This still didn't deter most of the fairgoers, who were willing to fork over cash to look at the newfangled machines and the tiny premature babies. The admission price that Lion had intended as a crowd-control method quickly turned into a viable funding source to build more of the lifesaving machines.

Another pediatrician, Dr. Martin Couney, joined the exhibition craze and also began exhibiting incubators with preemies inside. Couney's exhibit was so successful that he was asked to exhibit all over the world–which he did, finally ending up at Coney Island's Luna Park, his first permanent exhibit, in 1904. The sign at the show read ALL THE WORLD LOVES A BABY, and it became Coney Island's longest-running show. It also saved a lot of lives–New York hospitals began routinely sending all premature babies to Dr. Couney.

The babies received this excellent around-the-clock care for free, and their families were given free passes to the exhibit. The results were miraculous. According to statistics at the time, of premature babies born without the use of an incubator, only 15 percent lived. With the use of the incubator, 85 percent survived. More than 6,500 of the 8,000 premature babies used in the Coney Island exhibit survived and were sent home to their families.

Couney kept his exhibit going for many decades, until the rest of the medical world finally caught up with the incubator sideshows and began opening hospital preemie centers of their own.

Doctor, Please Don't Humour Me

Why was bloodletting thought to heal the sick?

Long after the Ottomans, the Chinese, and the Byzantines had realized the finer points of anatomy and physiology, western Europe stayed stuck in an old paradigm they'd learned from the ancient Greeks. It went something like this: The body consists of the four humours (fluids): blood, yellow bile, phlegm, and black bile. For the body, mind, and character to be healthy, these four fluids need to be in perfect balance. If something is wrong in the body, it must be the result of imbalanced humours.

Not just the body was affected by the four humours, but the personality as well. If someone were lecherous, warlike, and rash, they clearly had too much blood. Anger, violence, or vengefulness could be attributed to an excess of yellow bile. Cowardice, paleness, or dullness came from too much phlegm, and laziness, an overwrought disposition, or gluttony was blamed on too much black bile.

Common medical belief held that a good diet, exercise, and a good environment would keep the four humours in balance. However, if they got out of whack, there were plenty of noninvasive remedies: laxatives or diuretics, hot baths with herbs, smoke from burning herbs, and the like. For more serious situations, though, there was bloodletting. The reasoning behind the procedure was that if blood amounts were lowered, the other fluids would fall into balance. It was a crazy idea—and it didn't work. Bloodletting killed more people than it cured, yet it was performed on the sick well into the nineteenth century.

Electronic Gadgets

They amuse us, inform us, connect us to distant places, think for us, and tell us where we're going. Meanwhile, nobody is quite sure whether electronic devices are improving humans or degrading us, making us smarter or stupider, so let's examine a few more closely and see if we can make sense of it all.

Watson's Not Available, Please Leave a Message

What were the first words ever spoken over the telephone?
As with so many great events in history, the first words spoken over the phone came purely by accident. When

Alexander Graham Bell spilled acid on his pants in his workroom, his shriek of pain and panic came through a system he was testing. In the next room, Thomas Watson heard Bell's electronically transmitted voice shouting, "Watson, come here, I want you!"

What else did Alexander Graham Bell invent besides the phone?

A man-carrying kite, a land-mine detector, the hydrofoil boat, takeoff and landing gear for the airplane, and the aileron–a part of an airplane wing that moves and helps control rolling. He was also a cofounder of the National Geographic Society.

Phoneless Cords

Do tin-can telephones really work?

They do, actually. The concept is pretty basic: When a person places her mouth inside the opening of a can and talks, the sound from her voice vibrates against the bottom of the can. These vibrations travel down the string and make the bottom of the other person's can vibrate, too, transmitting the speaker's voice into the listener's ear.

What works even better than a tin can is a paper cup. It has less resistance, so the sound vibrates even better. But it's imperative that the string be pulled tightly all the way across, or it won't vibrate well enough to transmit sound.

This concept is actually very similar to the way a telephone works. While a telephone uses electricity to transmit the sound vibrations, the principle is the same. It's just as we always thought: play *is* educational.

Back to Your Cell

Do cell phones cause brain cancer?

Maybe. Probably not. Scientists aren't completely sure. It's fairly accepted that cell phones don't cause cancer with short-term use, but since cell phones haven't been around long enough to be considered "long term," it's too early to say for sure.

That said, major studies in recent years indicate that regular cell phone use (at least sixty minutes a day for four or five years) does not lead to a higher incidence of brain cancer, nor is there a higher incidence of cancer along the side of the head where the phone is used the most. So far, the biggest danger of cell phones seems to be when users insist on driving while using them (see below). That, and the violence that threatens to flare up when users yak on their cell phones in theaters, meetings, and restaurants.

Shut Up & Drive

Do more accidents happen when people drive while talking on their cell phones?

Yes. Basic common sense dictates that driving while mentally focused on a conversation slows down reaction time by as much as half a second. But why take our word for it? Here's what a major study on the topic found:

- If you talk on a cell phone while driving, you're four times more likely to get into an accident than if you weren't using your phone.
- After you hang up the phone, your chances of having an accident are pretty high for about fifteen or more minutes.
- Although age, gender, and driving experience don't seem to affect the odds, younger people tend to have a slightly higher risk of accidents while gabbing on their cell phones than older people. But this age-to-accident difference is consistent with non-cell-phone-using drivers as well.
- Researchers also discovered that it doesn't seem to matter if your hands are free or not—whether you're using a handheld phone or a hands-free car phone, the risk is close to exactly the same.
- If you use a cell phone, you have a greater chance of causing an accident while traveling at a higher speed than when traveling at a slower speed (in, for example, a parking lot).

Furthermore, investigative reporters for the *Portland Oregonian* found that cell phone towers interfere with emergency vehicle radio signals. Twenty-one of the twenty-eight states studied reported hundreds of incidences where emergency radio signals were cut off, intercepted, or otherwise disrupted by signals from cell phone company towers. Hang up, America, and drive—it's okay to be out of reach every now and again.

Communicating with Flare

What are sun flares, and why do computer networks and phones go out during them?

Sun flares are massive explosions on the sun with enough force to burn through the upper layer of Earth's protective atmosphere. The flares are magnetic, so Earth's magnetic field is able to protect us from the huge amount of heat and energy. However, the flares can still manage to disrupt activity on Earth's surface for minutes or hours at a time, by scrambling radio, TV, phone, and other signals. But it's not just modern human stuff that's affected—the jolt of magnetism is enough to throw off the navigational ability of some birds and sea animals and disrupt the growth of some plants.

Musical Laser Beams

How long is the groove on a CD?

It depends on how the cats are blowing, daddy-o. (Oh, sorry—we momentarily flashed back to our neo-Beatnik days.)

Anyway, the groove on a compact disc isn't exactly a groove in the same sense as what's on a phonograph record. It's more like a path of binary-coded bumps (which some people call "pits," though that's a misnomer). The bumps are so small that only a laser beam could find them. The laser beam reflects off the bumps differently than it does off the flat parts in between, shining back into a sensor that interprets the flashing reflections as shining either "on" or "off." The electronics of your CD player interpret this as a series of ones or zeros,

44,000 times a second, and decodes these ones and zeros into music.

Not surprisingly, it takes a lot of bumps to convey all of this information to your CD player and speakers. That's why the total distance covered by the laser beam when playing a CD is more than three miles.

Where did the name laser come from?

Like sonar and radar, laser is an acronym for a phrase that is much longer. It stands for "light amplification by the stimulated emission of radiation."

How many RPMs does a CD spin?

It depends. The old phonograph records spun at a constant speed (for example, 33 or 45 revolutions per minute). That made it easy to make a turntable to handle them, but it meant that the music recorded near the outside of the record had better sound than the music recorded on the inner grooves.

A CD solves this by varying its speed as it plays, so that the speed of the bumps below the laser beam always stays constant. That means that as the CD begins playing, the music on the inside groove (yes, the album starts on the inside and works its way toward the outside) sounds as good as the music near the outside. On the inside grooves, a CD spins at a rate of 200 revolutions per minute and gets as fast as 500 rpm as the laser beam approaches the outside edge.

DVD Killed the Video Star

How do they get so much more information on a DVD than on a CD?

It's a challenge, all right. DVDs use similar material and manufacturing processes, but there are several differences. Part of the solution was making the bumps that carry data smaller and jamming them closer together, making the "groove" of a DVD 7.5 miles long, more than twice what you get on a CD. But wait, there's more. The designers managed to double that capacity to 15 miles per side, because the DVD uses a sandwich of two different levels of bumps. Like a CD, the DVD has a reflective aluminum base behind the inner layer. However, on top of that is a semireflective gold layer, which also contains data bumps. The result is that the laser can read the top layer, then focus through it and read the inner layer as well. The cumulative effect is that a typical DVD has 7.5 gigabytes of storage.

What does DVD stand for?
Digital video disc.

Why can't they record on both sides of a DVD?
They can, they just usually choose not to. Manufacturers believe that consumers would rather have a colorful label printed on one side than have a two-sided disk with twice the capacity.

Still in the Sky

How do they get communications satellites to stay in one place in the sky?
The key, of course, is to get the satellite into an orbit that is exactly the same speed as Earth (one time around every twenty-four hours). A space region called the Clarke Belt, about 22,300 miles above the surface of the planet, is the place where you find a large number of broadcast, weather, and communications satellites. The Clarke Belt is named after science-fiction author Arthur C. Clarke, who back in 1945 first suggested that communications satellites could travel at the same speed that the globe spun, making them appear to hang stationary over one spot.

And This One Goes Out to Radioman Fred & the Boys

Wasn't Marconi the first person to transmit voices by radio? A newspaper column said it was a Canadian guy.
Italian Guglielmo Marconi was the first to send wireless transmissions, but his broadcasts contained Morse code, not voices. We're not saying that his accomplishment wasn't a big deal—by 1901, for the first time in history, messages could be sent instantaneously across the ocean to Europe. Radio transmitter/

receivers quickly became standard equipment on ocean vessels, and the "radio man" on ships became an important position.

Fast-forward a few years to Christmas Eve 1906. Imagine being a homesick radio operator in the middle of the Atlantic Ocean, keyboarding Morse code messages of seasonal cheer to other homesick radio men, and to friends and family on the mainland. Suddenly, among the familiar dots and dashes, a voice came upon the midnight clear: "Merry Christmas!" It was Reginald Fessenden, a Canadian engineer who had once worked for Edison's studios, broadcasting from his experimental station in Massachusetts. He gave a short speech, put on a record of Handel's "Largo," played "O Holy Night" on violin, sang some Christmas carols with his wife, Helen, and wished all the astounded radio men at sea a wonderful holiday and a good night.

Few people had shared Fessenden's belief that broadcasting voices was possible. When he asked the opinion of a former employer, Thomas Edison, Edison replied, "Fezzie, what do you say are man's chances of jumping over the moon? I think one is as likely as the other." Fessenden had made some short experimental broadcasts earlier using microphones, but none as dramatic as his Christmas Eve special. Although Fessenden's work proved that voice radio was possible, it would take another decade and the needs of the military in World War I before voice broadcasts became a commonplace thing.

Music from the Soul of Science

When was the first electronic music synthesizer built?

In 1906 Thaddeus Cahill first unveiled the Telharmonium, an instrument made up of 145 modified electrical generators and specially geared shafts and inductors that produced alternating currents of different frequencies. Since it was twenty years before the invention of the amplifier, Cahill ran the resulting current through the Bell Telephone system to phone receivers equipped with accoustical horns (like those found on old Victrolas). Each instrument was huge in size, and it severely disrupted local phone service, so the Telharmonium didn't get far beyond one favorable review by Samuel (Mark Twain) Clemens, a man who loved gadgets of all sorts. Still, its basic generation-motor-and-gear technology turned up again in the 1930s inside the wildly popular Hammond Organ.

What instrument makes that unearthly "Whoo-EEE-OOOO" sound in the sound track of horror movies?

It's a theremin, which sounds like it should be one of the B vitamins, but instead it's a spooky-sounding electronic instrument developed by a Russian scientist named Leon Theremin in 1920. The instrument used two radio wave oscillators to generate the sounds, and musicians obtained its sliding sound by moving their hands along the theremin's antennae.

Besides being found mostly on the sound track of horror movies, the theremin's greatest hit was "Good Vibrations" by the Beach Boys.

Who or what was the Moog synthesizer named after?

The answer is Dr. Robert Moog, the man who invented the instrument that helped define the psychedelic sound of the late 1960s. A decade earlier, as a graduate student at Cornell, Moog financed his musical science dreams by selling about a thousand of his do-it-yourself Theremin kits for $49.95 each, banking profits of $13,000 (the equivalent of about $110,000 in today's money). In 1965, he unveiled his first Moog synthesizer, an unwieldy, room-sized contraption that could play only one note at a time. Luckily, though, that was enough—thanks to the newfangled multitrack recording machines of the time, record producers could overdub note upon note to create an overall symphonic effect. With time, Moog created synthesizers that were portable, less expensive, and actually able to generate chords.

> Where can I hear sound samples of theremins and Moog synthesizers?

How does a GPS work?

A global positioning system receiver is a pretty cool little gadget. It's somewhat based on the navigational systems of old in which skilled sailors using compasses, astrolabes, and sextants could tell where they were (give or take a hundred miles) by combining the data of the stars, sun, and moon positions. Of course, there were problems. It took a lot of math skills, charts, and instruments; you couldn't see the stars except on a clear night; and the precision left a lot to be desired.

A GPS receiver bypasses all of the problems by using satellites, radio waves, and sophisticated electronics. The United States uses twenty-four orbiting satellites to ensure that wherever you are in the world, there are at least four satellites your GPS receiver can pick up (three satellites is the bare minimum needed to locate your latitude and longitude; if you also want to know your altitude, you need a fourth one as well).

Each satellite broadcasts an identifying signal at the speed of light, which includes time information that allows your GPS to figure out how long it took the signal to arrive, which then allows your GPS to figure out how far away each satellite is. By comparing these distances from several satellites, the GPS can figure out your exact location—to within a few feet.

How do you use a GPS to play the hide and seek game of geocaching?

Just the Fax, Ma'am

I saw my first fax in the 1980s, but my friend swears they've been around forever. True?

Longer than the telephone, as a matter of fact. A Scottish clockmaker and inventor named Alexander Bain patented the first facsimile machine on May 27, 1843–thirty-three years before Alexander Graham Bell's invention.

It took another few decades before the invention got any significant use. A commercial fax service (called "pantélégraphes") opened between the French cities of Paris and Lyon in 1865, but faxes didn't really come into their own until 1906, when newspapers began regularly running photos in their pages and a device that could transmit photo images over long-distance phone lines became a necessity.

For more than seventy years, news photos were the primary use for fax machines, but that began changing in the early 1980s. Ironically, you can credit Federal Express for a near-universal adoption of fax machines, while being perhaps their biggest victim as well. FedEx helped whet the appetite for faster mail in the first place with the introduction of its overnight mail service.

When its management team got wind that several manufacturers were developing fax machines for business use, FedEx panicked, figuring that fax could eliminate the need for most of their overnight express services. They ordered 13,000 fax machines, complete with customized purple ink to match their logo, and, beginning in 1984, heavily advertised their "Zap Mail" same-day service: for $3 to $4 a page, they would send a driver to pick up your document and take it back to the local office to zap it to another office, which would then deliver the facsimile to its destination . . . all within hours.

At the time, well before e-mail and when most people had no idea what a fax machine was, Zap Mail was considered magic. But shortly afterward, fax machines started flooding the market. It didn't take businesses long to figure out that it would be much cheaper and more convenient to buy their own machines than use Federal Express's expensive Zap Mail service. FedEx helped create a market, but when it dropped Zap Mail in 1988, it ended up taking a loss of $250 million when it got stuck holding 13,000 machines and barrels full of purple ink.

How does a fax machine work?

The technology is pretty simple. A bright light reflects off the document as it passes along a strip containing hundreds of tiny photocells. If a white area of the page passes below, the photocells convert the light shining off the document into electricity. If the area is black, the photocell generates no electricity. These impulses fire off in quick succession and are converted into sound signals that travel over the telephone lines. The fax on the other end "hears" the signals and interprets where the black should print. If it's a thermal fax, tiny pins, corresponding to the hundreds of photocells, heat up to match the black areas of the original document. The heat activates those sections of the heat-sensitive paper, turning it black as it passes by.

That's how most home fax machines work. However, there are other variations like plain paper fax machines, which use the same idea but with a laser or inkjet printer instead of heatable pins.

Not Polite to Point & Click

Why do they call the computer device a mouse?

Because, for some reason, the "X-Y Position Indicator for a Display System" just didn't catch on. That was the name given by the device's inventor, Douglas Engelbart. The first users hated the clumsiness of the name and quickly dubbed them "turtles," which eventually became "rodent," which morphed into the cuter-sounding "mouse." This name was just right for the shape and size of the X-Y Position Indicator, and it stuck.

Some Day My Prints Will Come

My computer came bundled with a "bubble printer." Does it really use bubbles to print?

A bubble jet printer uses a principle as old as mud bubbling out of the ground. Have you ever noticed that when bubbles pop, a little of the liquid gets propelled outward? Sure you have—think of the misty feeling you get when you put your nose up to ginger ale, or the stickiness that sometimes happens if you're too close to a gum bubble when it pops.

Anyway, popping bubbles inspired inkjet printer designers who wanted to propel a microscopic dot of ink onto a piece of paper quickly and on command without actually touching the paper. A bubble jet print head typically has 300 to 600 microscopic nozzles, and a tiny amount of ink waits inside each nozzle. As the print head quickly stops and starts along the page (so quickly that it seems like a continuous motion), jolts of electricity heat up a resistor in the nozzle. This heat instantaneously boils the ink, vaporizing it into a bubble and launching it from the print head to dot neatly onto the surface of the paper. How small are these neat little splatters? It can take dozens of them, precisely fired, to make up a single letter on the page.

Briefly, how does a laser printer work?

Briefly? We're not sure we know the meaning of the word. Especially not with something this complicated. But we'll try.

Inside the printer is a large metal roller (the "photoreceptive drum"). As the drum rotates past an electrified wire ("the corona"), the surface gets a positive electrical charge in preparation for the laser beam.

The beam is guided by the computer, of course, invisibly scanning the text and images onto the photoreceptive drum. The metal parts the laser hits build up a temporary negative electrical charge and immediately get brushed with toner powder, a mix of positively charged plastic and pigment dust. Because opposites attract, the positive toner sticks to the parts that had been zapped negatively by the laser, but is repelled by the rest of the drum, which is still positively charged.

At this instant, that part of the drum looks like a mirror image of the page being printed. But not for long, because it immediately makes contact with the paper. An electric wire under the paper (the "transfer release corona") zaps the paper with negative electricity, which pulls the powder off the roller and onto the page in the pattern drawn by the laser.

Its ink held on only by gravity at this point, the paper then passes between the "fusers," a pair of hot metal rollers. The heat melts the powder to the fibers of the paper and spits it out of the printer. This is why paper coming out of a laser printer is always so toasty warm.

Computer Engineer @ Work

Who invented e-mail?

A computer engineer named Ray Tomlinson. He's also the guy who designated the @ sign as a crucial part of e-mail addresses. Long before the Internet as we know it existed, there was ARPANET, an experimental forerunner commissioned by the Defense Department in 1971. It consisted of fifteen universities hooked together by a rudimentary network. That year, Tomlinson, accustomed to sending electronic memos within his own university, wondered why he couldn't do the same thing to other users in the network.

But how to indicate that the mail was supposed to go to another specific institution? He thought for a few minutes and then chose the @ symbol to distinguish messages that were headed out onto the network. "The @ sign seemed to make sense," he recalled later, appropriately in an interview conducted over e-mail. "I used the @ sign to indicate that the user was 'at' some other host rather than being local."

His first e-mail message? Nothing as memorable as Samuel Morse's "What has God wrought?" or even "Watson, come here, I want you." In fact, the first e-mail message was addressed to himself at the other computer in the same room. "I sent a number of test messages to myself from one machine to the other," he admitted years later. "The test messages were entirely forgettable: QWERTYIOP or something similar."

Fly Me!

Why and how did some computer nerd coin the term joystick?

The joystick predates computers by decades. It has been a part of airplanes almost since their introduction–its first known reference dates back to 1914. An airplane joystick works much the same as with computer games–you pull or push it, or lean it to the side, to control direction of movement.

Where did it get its name? The reference books don't seem to be sure whether it's a sexual reference or not. We do know that at some time or another a "joystick" was an opium pipe, but whether that's related to its aviation usage is not completely clear. One aviation source claims that it was invented by "a man named Joyce" and that the name *Joyce stick* got misheard and shortened to *joystick*. (On the other hand, most aviation historians credit French flier Robert Esnault-Pelterie–not someone named Joyce–for inventing the joystick in 1907.)

We do know that the joystick terminology got one of the early pioneers of personal computers in trouble with customs officials back in the 1970s. He went to the airport to receive a shipment of boxes marked "Joysticks" and was taken into custody in the belief that the odd-looking plastic devices with the lurid name were some sort of exotic sexual aid.

Polaroid Schizophrenic

How can Polaroid instant photos develop before your eyes? How come the light outside the camera doesn't instantly overexpose the photo?

Polaroid photos are seemingly a magic trick, and in reality there is something of an illusion in the way it "develops before your eyes."

After you take an instant photo, it passes through two stainless-steel rollers, which spread chemicals that are collected in a blob at the edge of the plastic film sheet. These chemicals start developing the photo, but the rollers also do something you wouldn't likely imagine—they spread a layer of "opacifier," which is a thick, white, opaque coating that keeps light from reaching the photosensitive layers below.

At this point, the photo underneath has already developed, yet you still can't see it. That's because the opaque layer is still in place. It's only after the photo is completely developed below that acid seeps up and reacts with the opacifiers that cover it. When that happens, the opaque coating slowly becomes translucent, then transparent, slowly revealing the photo that's already fully developed below. So while it looks like the photo's developing before your eyes, all that's really happening is that a layer that keeps you from seeing the photo is slowly disappearing.

Bloop-Bloop

How does a lava lamp work?

A lava lamp works on two principles: heat rises, and some liquids don't mix. Waxy liquid shares a glass bottle with thin, watery liquid. The bottom of the metal base underneath the glass bottle holds a light. When you plug it in and turn it on, the light slowly heats up the contents of the bottle. After a while the waxy blobs melt, and these liquidy glops rise with heat. At the top of the bottle, they cool and drop back to the bottom to be reheated again.

What the stuff in the bottle actually is remains a secret. The companies that produce the lamps won't say, and chemists have come up with slightly different results when analyzing the stuff. What they do know, though, is that the waxy stuff–the lava–is probably paraffin, with perhaps a stabilizer that helps keep the blobs together when heated and moving. The colored "water" might be just that–colored water–although most agree it's probably an alcohol-based liquid that remains stable even when heated.

How can I make my own lava lamp?

A Critical Mass of Miscellany

So what does one do with all the extra questions about the wonders of science? Why, we put them in a chapter all their own in the hope that their gravity will pull you through to the end of the book.

Larry, Moe, & Kirlian

What's with that photography that's supposed to measure your aura or your "bioenergy"? How is it done, and is there anything to it?

It's called Kirlian photography, and there is indeed something to it. However, it isn't what its parapsychological proponents claim.

The process is named after Semyon and Valentina Kirlian, a husband-and-wife team who first reported their findings in the *Russian Journal of Scientific and Applied Photography* in 1961. What they discovered was that if you put something that is or was alive on photographic paper and run a high-voltage electrical charge across it, the resulting photo will show a glowing, colorful "aura" encircling it.

There's no doubt that something is showing up. However, the question is what it is. It's not completely understood, but it seems to be due to natural phenomena such as electrical grounding, humidity, air pressure, and temperature. Changes in any of these things will produce different "auras."

How about the most impressive trick—tearing a leaf in half and having the missing half mysteriously show up in its Kirlian image? Well, the effect almost never shows up spontaneously. If you want to make it a surer thing, here's how: Make sure the leaf is moist. Set it down and take a photo, then tear it in half and carefully place the leaf back into the same position. The residual moisture may give you the desired mystifying effect.

Colors of a Sickly Hue

Why did Francisco Goya's work change so drastically from conventional portraits to dark and grotesque subjects?

It was probably the result of lead poisoning. Through his mid-forties, Francisco Goya was not only a court painter but also well known for his tame and innocuous subjects. But all that changed when he was forty-six. In 1792 he contracted a strange and mysterious illness that left him incapacitated. His vision went blurry, he suffered from coma and partial paralysis, and he struggled with bouts of dizziness and hearing loss. Most important to his artwork, however, was a sudden onset of paranoia and hallucinations.

The sickness had a dramatic effect on his paintings. His images turned from light and airy to dark and tortured. His nightmarish scenes stirred up great controversy.

He recovered with time, but what caused this weird illness? Many historians agree that, because he was one of the artists who mixed all of his own paints, he probably developed a very serious case of lead poisoning (and possibly poisoning by other ingredients as well). Most artists of the time mixed their own pigments using lead, cadmium, mercury, and other deadly substances. Goya, in particular, was known for his messy way of painting—with brush, trowel, rag, and hands furiously moving as he mixed and worked, likely absorbing the toxins through his lungs, mouth, and skin.

Throughout the rest of his career, Goya would repeatedly suffer bouts of the same mysterious sickness. When he fell ill, he was forced to quit working. That fact might've saved his life, because his layoffs would've allowed the lead levels in his blood to drop—at least until he felt good enough to start working again, starting the cycle all over again.

Some art historians say the poisoning may have actually helped his career considerably. He developed a reputation that would follow him long after his death—a gift that his earlier, more subdued paintings would never have given him.

Why did the British government ban the Indian yellow hue of paint in 1908?

Indian yellow was first introduced in the 1750s and quickly became a staple of every artist's palette because of its rich yellow hue. To get the deep color, manufacturers fed mango leaves to cows, collected their urine, and made a concentrate from it. In 1908 officials decided that feeding large quantities of mangoes to cows was inhumane, so the practice and the color were banned. However, it wasn't until about fifty years later that paint ingredients more dangerous to humans were banned (see question above). Some cynics believe this suggests that the value of a cow is greater than the value of an artist to the British government.

The Write Stuff

Who invented the typewriter?

There were many patents for typewriterlike machines, but only in 1868 did Christopher Latham Sholes invent the first typewriter that was also practical to use. The Remington Arms Company began producing it in 1873.

One of its biggest fans was writer/humorist Mark Twain. He called it "a curiosity-breeding little joker" and was one of the first people to own one. As a result, in 1883 he became the first author to ever submit a typewritten manuscript to a publisher–his book of riverboat memoirs, *Life on the Mississippi.*

What's graphite, exactly?

It's a rock–a carbon compound that's been used for writing since long before modern pencils were invented. But it was an eighteenth-century geologist–Abraham Werner–who named the stuff *graphite,* from the Greek word meaning "write."

If pencils don't have lead in them, then why are they called "leads"?

Seeing that pencil "leads" are made from graphite, the name doesn't make much sense, does it? But there once was a time when people thought graphite was a type of lead, hence the name.

Random Notes & Half-Notes

I'm tone-deaf and can't sing on key. Is there a way to undo that?

First of all, very few people are actually, clinically tone-deaf. It means they have a neurological condition that renders them totally unable to hear differences in pitch. People who are truly tone-deaf can't even hear the pitch differences in speech–for example, whether someone's voice is going up when asking a question, or staying even while making a statement. Do you have trouble discerning the different emotions in people's voices when they're happy, sad, or goofing around? If you're okay in that department, the good news is that you're not technically tone-deaf, and there's a lot you can do to improve your singing.

Most people who can't sing on pitch are simply suffering from a lack of training and practice. Some people need more exposure to music than others to master the ability to discern and then hit the right notes. While for some of them it may be a difficult process to learn, it can be done.

What's the difference in frequency between middle C and the Cs an octave below and above?

Middle C is vibrating at a rate of 261.625 hertz. The C above middle C is vibrating at 523.25. The C below middle C is vibrating at 130.8125. You may notice that the higher C is vibrating exactly twice as fast as middle C. The lower C is vibrating at half the speed. This rule can be applied to any two notes separated by an octave.

How high does a sound have to go before its frequency will damage the human ear?

Loudness hurts the ear, but frequency doesn't. Once it goes above a certain frequency, the human ear simply won't hear it.

Cotton Candy Blend

I've seen polyester chips, which are hard little pieces of plastic. But I can't figure out how they make them into cloth.

Have you ever watched a cotton candy machine? You put colored sugar granules in the top of a hot metal cup that has tiny holes in it. The sugar melts, and when the cup rotates at a high speed, tiny threads of molten sugar come shooting out of the spinnerets (the little holes) and, when they hit the cool air, immediately harden into sweet cobwebby material. The machine's operator gathers the spun sugar around a cardboard cone and sells it at an inflated price, considering that it's basically just a few tablespoons of sugar.

Now pretend that the polyester chips you saw are sugar, and essentially repeat the process above, but with a larger machine. That's how they make the basic polyester fiber. After being spun, it's usually wrapped around a spool, heated again, stretched, and then wound into threads. During that process, polyester is often mixed with other fibers, notably cotton. While polyester lessens wrinkling, the addition of the natural fibers lets perspiration absorb away from the body, solving the affliction of the "Disco Fever Sweats" that plagued wearers of 100 percent polyester clothes in the 1970s.

Stuck on You

How much Velcro would it take to hold a person on a wall?

Given the fuzzy pajamas that babies are wont to wear, we've often thought that panels of Velcro around the house would be a great way to get them off the floor and out of the way. Unfortunately, the good people at Child Protective Services had other opinions.

But an adult? Let's consider David Letterman as our prototype, since he's the one who popularized the concept many years ago, and assume he weighs about 175 pounds. According to a Caltech analysis, the best grade of Velcro (yes, there are grades) will hold from 10 to 14 pounds for every square inch of Velcro. If you can figure out a way to firmly and completely place your body mass against the wall, you could theoretically get by with as little as thirteen square inches of Velcro–smaller than a square four inches by four inches. Unfortunately, that's a theoretical number that will only work when laying one flat surface against another. Since your body is irregular and rounded, we'd spring for more Velcro if we were you.

Who invented Velcro, anyway?

A Swiss engineer named George de Mestral. In 1948 he was walking through woods and fields with his dog, presumably yodeling into the mountains, when he got intrigued by cockleburs that were getting stuck to his pants and his dog's coat. Being an engineer, he took some home, looked at them under a microscope, and determined that J-shaped hooks are what kept the cockleburs stuck. He figured out how to make some similar tiny shapes with plastic, and added a fabric with plenty of loops for the hooks to grab onto. He named it "Velcro" from two French words: *velour,* "velvet," and *crochet,* "hook." The results are something we've been stuck with ever since.

Ask Us about Volcano Weather Next

What's "earthquake weather," and how can I use it to predict The Big One?

Earthquake weather means different things to different people. Some people believe that long stretches of abnormally hot temperatures precede earthquakes. Others believe bouts of low humidity are good indicators. There are as many weather theories as there are weather patterns. The truth is that weather plays no apparent role at all in the shifting of Earth's plates.

Deep and Wet

Where is the lowest place in the United States?

Death Valley, California, is situated 282 feet below sea level, making it the lowest dry land in the United States.

Where is most of the world's fresh water?

About 90 percent of the world's total fresh water is frozen in Antarctica.

Like a Rock

What kind of rock is the Rock of Gibraltar?

Gibraltar stands as the northern gateway to the Mediterranean Sea, and has historically been a much-coveted fortress for observing all the movement in and out of the area. Despite its history as an impenetrable fortress, the Rock of Gibraltar consists almost entirely of soft gray limestone with a touch of shale thrown in for good measure. Furthermore, it's full of holes—more than 180 caves carved by wind and rain in the deteriorating limestone. We suppose it goes to show that "solid as the Rock of Gibraltar" is not so solid after all.

I've got a geology teacher who claims that acid rock, country rock, and hard rock were all geological terms before they were names for types of music genres. Is that true?

He forgot *boss rock*, which was a geological term long before Bruce Springsteen came along. But seriously, yes, your teacher is correct. If you're looking for a good genre name, it's always best to turn to science first. Check out some of these other geological terms and see if any could apply to your band's music:

> rudaceous rock
> pryoclastic rock
> grit rock
> gangue rock
> boudinage rock
> arenaceous rock
> conglomerate rock
> dyke rock
> porphyritic rock
> fractured rock

One exception we have to note, however: the term *heavy metal* comes not from the field of geology, but from chemistry.

Money Matters

So how much copper is actually in today's penny?

Only about 2.5 percent of a penny is copper. The layer is just thick enough to give the coin its distinctive copper color. Most of the penny is made with zinc, but that's a relatively new development in the history of the coin. Before 1982, pennies were about 95 percent copper. So why the switch? In the early 1980s, the amount of copper included in a penny became worth more than one cent. Facing rising costs and the specter of pennies being melted for their metal, the mint chose a cheaper metal instead.

What was a shekel made of?

Silver. One shekel weighed anywhere between 8.26 and 14.5 grams, depending on when it was made (and which historical source you ask).

Is there a name for the fear of money?

Yes. If you have a deep-seated fear of money, you're suffering from chrematophobia.

Going the Extra Eighteenth of a Mile

The distance of a mile doesn't seem to make sense. Why is a mile 5,280 feet instead of an even number like 5,000?

Blame it on the Tudors and their fondness for the furlong. The mile was originally 5,000 feet, but Queen Elizabeth I wanted it divisible by a furlong. Each furlong measured exactly 220 yards, so in 1575 she made it a royal decree that the mile would officially grow 280 more feet, making it exactly 8 furlongs.

Why is a marathon 26.2 miles? Wouldn't it make more sense for it to be an even 26 miles?

Well, yes, but British royalty (as noted above) don't always do things because they make sense. The original marathon from the first modern Olympics in 1896 was actually based on physical landmarks. It began at Marathon Bridge and ended at Olympic Stadium in Athens. This gave it a distance of 24.85 miles, or an exact 40 kilometers. This made sense until the 1908 Olympics, when the race ran from Windsor Castle to White City Stadium in England—a distance of exactly 26 miles. However, the British royal family decided the race should finish right in front of their viewing box. Unfortunately, getting the runners to end at that point meant tacking on an unnecessary 385 yards. The royal family seemed to think it all made perfect sense and put their royal feet down. For whatever reason, this odd distance of 26 miles, 385 yards stuck as the official marathon distance thereafter and to this day.

Not Even the Wall of China

What's the only man-made object visible from the moon?
You can't see *any* man-made object from the moon. If you look
at pictures of astronauts on the moon with the full Earth
behind them, you can see why–Earth from the moon doesn't
look much bigger than the moon looks from Earth. But we
don't fault you for thinking otherwise, because this myth has
been around since before *Sputnik* was launched into space.

The only objects on Earth visible to the naked eye from the
moon are water, clouds, and sometimes some of the land
masses, if the clouds don't get in the way.

To Infinity and Beyond

***If there's no air to push against, why do rockets work in
space?***
It's a good question. You'd think that a rocket's exhaust has to
push against *something*, like a paddle needs to push against
water to propel a canoe.

But in reality, a rocket doesn't need to push against the
atmosphere to fly. What it's doing is essentially pushing against
the inertia of the rocket exhaust. But that's confusing. Let's try
to look at it from a more earthbound angle.

Let's say you're sitting in a shopping cart with a bowling
ball in your hands. You lift the bowling ball over your head and
toss it as hard as you can toward the back of the cart. If you're
on smooth pavement with well-oiled wheels, your cart is going
to move forward at a good pace.

If someone hands you bowling ball after bowling ball for
you to toss in quick succession, you'll be able to keep your
cart moving forward. You've created the equivalent of a slow-
motion rocket engine. But remember that the reason you're
moving is not because the bowling balls are pushing against
the air–it's your action of tossing them that's pushing the
cart forward. That would be true even in the vacuum of
space.

What happens with rocket engines is that as rockets burn
fuel, they release a huge collection of gases out the back at

a tremendous speed. The effect in forward motion is the equivalent of chucking thousands of bowling balls a second.

The Squids Are All Right

Do any animals use jet propulsion to get around?

The squid. It fills up a sac within its body with water and then forces it out through a tube below its head. This jet action makes the animals glide through the water at pretty high speeds (see page 147).

Nearer My Dog to Thee

What happened to the little Russian dog that was the first living thing in space?

On November 3, 1957, less than a month after its first ever space launch, the Soviets sent up *Sputnik 2,* complete with a dog named Laika. She didn't survive in space for very long. Conflicting accounts put the number of days at somewhere between four and seven, but that was it. The Soviets, rushing to beat the Americans in launching an animal into space, had made no provision for bringing Laika down safely, and so she died when the oxygen supply ran out.

She's not still up there, frozen solid, though. Her spaceship reentered Earth's atmosphere five months later, and Laika burned up in a blaze of glory.

A Light Dinner Indeed

I saw a picture of an astronaut eating and drinking from an open container in space. Wouldn't the liquid float out?

Surprisingly, eating and drinking in space isn't quite like you've seen in movies. Not everything has to be sucked out of a tube. It turns out that astronauts can eat most foods normally without having them float out of their containers. True, food containers need to be fastened to a surface with

Velcro, magnets, or sticky tape. But once they're there, foods and drinks will usually behave somewhat normally because of surface tension, as long as the astronaut eats slowly and gracefully. A liquid in a cup will tend to act like very thick ketchup, sticking to itself and the cup, unless dislodged with a firm jolt; most foods come with gravies or sauces and will stick to plates, forks, and spoons unless jarred or flung.

It's Da Bomb

How many nuclear bombs have gone off since World War II?

It's hard to get an exact figure, because governments try to keep these things secret. We do know that up until 1996, there were 2,036 known nuclear bomb explosions. Three-quarters of those were exploded underground, with 511 taking place in the open air. More than 80 percent of the fallout from these explosions is still circulating in the atmosphere; the rest has settled to the ground. Some scientists believe this radiation has been the cause of nearly 1.5 million extra cases of fatal cancer since the tests started.

Who Says It Isn't Rocket Science?

Were the Nazi rockets that bombed London the first time these weapons were used in warfare?

If that were true, how would you explain the "rockets' red glare, the bombs bursting in air," written during the War of 1812?

Now go back even further, to A.D. 1232. That year signaled the oldest known reference to using rockets in battle. At the time, the Chinese were fighting against Mongol invaders. The written record shows that the Chinese, experts at gunpowder, repelled the Mongols with a torrent of "arrows of flying fire." This arrow was a simple solid-propellant rocket—a tube of gunpowder attached to a long, pointed stick. When launched, they headed more or less in the direction they were pointed. The rockets weren't particularly accurate, but they were cheap to make, and could be launched in large quantities over a long distance. The actual physical effects were bad enough, but equally powerful were the psychological effects of hissing, flaming arrows flying into the Mongol ranks. The Mongol armies fled in panic.

What exactly were the "rockets" mentioned in the national anthem?

The British were pounding Fort McHenry with a weapon that had been used against them twenty years earlier while they were conquering India in the 1790s. Indian gunpowder experts had enlisted in the fight against British imperialism, pounding the British troops with a hail of primitive exploding rockets. Though India eventually lost, the devastating effect their rocket barrages had on British soldiers caught the interest of an artillery expert, Colonel William Congreve.

Congreve's rockets were inaccurate (people hadn't yet figured out how to use stabilizing fins and vanes), and they exploded when the fuse fire reached them, not on impact. The result was that a number of rockets went off in midair. So, like the Chinese and Indians before them, the British made up for the low quality of their rockets by increasing the quantity. Which makes sense, since "bombs bursting in air" have a terrifying effect, even if they don't land exactly where they've been aimed. From his vantage point as a prisoner on a British warship, Francis Scott Key likely saw thousands of these rockets launched against Fort McHenry. No wonder he was so worried he had to write a song.

A Lash of Thunder

What is making the sound when a whip is cracked?

It's the tip of the whip being propelled at over 700 miles per hour. When flicked just right, the end breaks the sound barrier, creating a small sonic boom. The sonic boom is what you hear when it's "cracked."

Land before Time

Which culture invented time?

No one invented time; it was likely discovered, even if subconsciously, by every culture. At some point early on, humans were aware of time passing and began keeping track of it. Although we don't know who was the first, archaeologists are discovering more and more ancient civilizations that devised their own unique ways of marking time. Stonehenge is an example. The earliest clock—a shadow clock similar to a sundial—can be dated all the way back to 3500 B.C., and the hourglass came into being not long after. Over time, humans invented the mechanical clock, then the pendulum clock.

Today we use a twenty-four-hour clock and have the world divided into time zones. The zones start at zero at the original site of the Greenwich Observatory in Greenwich, England. We go by what is called the Scientific Standard of Time, based on the second, which is defined by scientists as "the duration of 9,192,631,770 periods of radiation," and in other circles as "one Mississippi. . . ."

All Wound Up

How do self-winding watches wind themselves?

By using a little pendulum that swings when you move your arm. A few hours of natural arm movement will wind a self-winding watch enough that it can go for thirty-six hours without further winding.

You Go to My Head

Is there any wine that makes you drunk faster?

Yes. There is a type of wine that will make you tipsy faster than other wine. It's champagne. The little bubbles in champagne are carbon dioxide, which moves into your bloodstream faster. But there could be a more contextual reason why it gets you intoxicated more quickly than regular wine: it's usually consumed during momentous and celebratory occasions, allowing the giddiness of the event to add to the intoxication of the alcohol.

The Fat of the Land

How do they measure how many calories are in foods?

They use a hot little device called a bomb calorimeter. It burns foods and measures the difference between how much energy went in versus how much energy comes out. Here's where it gets confusing if you ever talk to a physicist about your weight loss program: What we call a "calorie" is actually equal to a *thousand* scientific calories, or a kilocalorie. So to a scientist, a doughnut doesn't actually have 235 calories, but 235 kilocalories (235,000 calories). You'd better start jogging.

How many calories can I burn while watching TV?

No matter how much high-impact channel surfing you might do with the remote control, you burn only 1 to 2 calories per minute.

The Science of Snack Food

How do popcorn kernels actually pop?

Not all corn kernels are equal. In order to pop, the kernel needs a water content of about 13.5 percent. Each popcorn kernel consists of soft, moist starch inside a hard outer shell. When heat is applied, the moisture expands and the starch is cooked. Eventually, the pressure gets so high that it bursts the outer shell with a loud *pop!*

What is the T.W.I.N.K.I.E.S. Project?

It stands for Tests with Inorganic Noxious Kakes in Extreme Situations, and was conducted by students at Rice University during finals week in 1995. These tests were conducted on Twinkies, with a special emphasis on experiments using the force of gravity, radiation, flame, and heat, etc. For more information, visit http://www.twinkiesproject.com/.

How long do you have to leave a tooth in Coke for it to completely dissolve?

A mighty long time. Despite the facts, this long-running myth just keeps going and going. A tooth won't dissolve in Coke or carbonated soda of any kind.

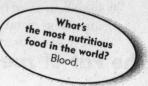

What's the most nutritious food in the world? Blood.

Don't Drop In Unannounced

If San Francisco's Golden Gate Bridge draws suicide jumpers, why don't they put up some fences to deter them?

It sounds reasonable, considering the bridge is open to pedestrians and is a popular tourist spot. The average is about twenty-two deaths a year. Only a few have ever survived the 250-foot drop into the chilly, raging Pacific below. We've heard that hitting the water from that height would not be a very peaceful way to go—the injuries jumpers incur before drowning are similar to hitting cement.

The city experimented with several deterrents, especially following a particularly bad suicide year in 1995. Some of these included a web of thin wire barriers, suicide prevention phones on the bridge, and antisuicide squads standing ready to answer desperate calls. But a fence has never really been an option, and not just because it would block the spectacular view. It would add so much wind resistance that in the strong Pacific winds the entire structure would be in danger of blowing over.

Metal of Freedom

Why is the Statue of Liberty colored green?

Burnished copper gives it that green hue. The copper sheets are no thicker than a penny, and were made in France, shipped over, then riveted to the iron frame of the statue. Because the rusted green copper actually serves as a coating that protects the integrity of the metal below, there have never been any attempts to clean it off.

Why is the tip of the Washington Monument made of metal, while the rest of the structure is made of stone?

When constructed in the 1800s, aluminum was a semiprecious metal, both hard to find and difficult to extract from bauxite (aluminum ore). So at the time, it was considered a fine luxury to have an aluminum cap crowning the top of the monument. Still, there were practical reasons as well. For starters, aluminum was safer than using stone because it was of lighter weight, adding less pressure to the entire structure. For another, the aluminum acts as a lightning rod. As a result, no other structure in Washington, D.C., is allowed by law to rival the Washington Monument in height. Considering the number of thunderstorms that rumble through the area, giving the structure a metal tip seemed like a mighty good choice.

The Weekend Scientist

This entire book you've taken our word for it. Now test out the principles of science for yourself, using these basic experiment guidelines. Then you can use your new-found knowledge to impress the heck out of your friends.

Many Cats Make Light Work

Can a lightbulb turn on in your mouth, like Uncle Fester's did in the TV show **The Addams Family?**

Not unless you're electrically charged. That, of course, would require having enough electricity running through you to electrocute you. Uncle Fester's lit-up bulb was just a stage trick for the show. You can, however, make a lightbulb glow without screwing it into a socket. Here's how:

Get a fluorescent lightbulb—any kind will work, but the more surface area, the better the results. Put on a nice, fluffy wool sweater, or find an incredibly good-natured cat. Quickly and forcefully rub the fluorescent bulb all over the wool sweater (or cat). The small amount of static electricity generated is enough to excite some of the gas molecules on the inside of the bulb. Find the Fester within, and watch it glow in an eerie way.

> **How can I make a pickle glow?** Ask

Waxing & Waning

I once saw a candle burning at both ends and rocking like a perpetual-motion drinking bird. Can you tell me how to replicate that trick?

Sure! Grab a candlestick, a match, and a pin, and let's do the impossible: burn the candle at both ends.

First, make sure you're working on a clean, flat surface like a bare kitchen counter or table—one that won't instantly go up in flames if the candle tips over, and one that also won't get ruined if splattered with wax droplets.

You'll need a needle, pin, or round, smooth toothpick; two sturdy ceramic or glass cups; and a lighter, along with your regular-sized candle (but *not* one of those dripless kinds). Cut or melt the wax at the bottom of the candle so that both ends have exposed wicks to burn. Stick your pin all the way through the center of the candle, perpendicularly and an equal distance from each end. Balance the ends of the pin between the cups, one end on each lip. The candle should be horizontal, dangling in the space between the two cups. Now light both ends and gently get it rocking.

It won't stop rocking as long as both flames stay lit.

What happens is this: Each time a drop of wax falls, the candle pulls up at the same time—it's a working model of Newton's law about an action creating an opposite and equal reaction. The weight of the candle also slightly shifts as that end suddenly gets lighter, helping to keep the candle rocking in a back-and-forth motion.

Rubbing You the Watt Way

What's the best way to generate static electricity?

Probably the best way to see if you've generated static electricity is to put on a pair of wool socks, run around on a plush carpet rubbing your feet briskly, and touch metal or another human. See how many times you come out unscathed. We also recommend a less painful way, like bending water. You'll need clean and dry hair, the bathroom sink, and a plastic comb.

First, turn the water on to a very slow, smooth trickle. Then comb your hair quickly about a dozen or so times. Very slowly move the comb close to the water stream, without touching the water. The water bends! Since the static electricity is creating a negative charge on the comb, it's attracting the positive charge in the water.

Another fun way to generate static electricity is with the Styrofoam often used to package meat and a lightweight metal pie tin.

First, cut a corner off the clean and dry Styrofoam tray and tape the corner to the middle of the inside of the pie tin. This is what you'll use to move the pie tin, so make sure it's taped firmly enough to be used as a handle. Now rub the rest of the Styrofoam tray on your head or on a wool sweater. Do this briskly, and then set it down on a nonmetallic table. Hold the pie tin centered six to twelve inches above the Styrofoam tray and carefully let it drop so it lands and stays on top of the tray. Don't let the pie tin come any closer than six inches or a foot to the tray before you let go, or the experiment won't work as well.

Turn out the lights and put your finger on, or close to, the pie tin for some cool fireworks. When the pie tin stops sparking, renew the static electricity by rubbing the tray briskly on your head or sweater again.

Get Out the Volt

How does the Life Saver spark trick work?

One of the three best in-the-dark revelations of adolescence is that if you crack a Wint-O-Green Life Saver between your teeth, tiny blue and green sparks will flash in your mouth.

If you've never seen it happen, pick up a pack of Wint-O-Greens and wait until after dark. Turn out the lights and crunch one between your teeth while standing in front of a mirror. Besides feeling silly for going to this much trouble, you'll see a blue-green spark.

But how does it happen? Nabisco says it gets about fifty to a hundred queries about this phenomenon every year—enough to have a canned answer: "Triboluminescence resulting from crystal fracture" at work.

Huh? Simply put, when crystalline molecules are smashed, the free electrons collide with the plentiful molecules of nitrogen in the air. The nitrogen molecules become excited and vibrate. They emit this extra energy in the form of mostly invisible ultraviolet light, but a small bit of visible light is emitted as well. This is the spark you see in the mirror.

All hard candies that contain sugar create triboluminescence when cracked. So why do Wint-O-Greens produce a visible result when other flavors don't? Because oil of wintergreen has the special ability to take in light with short wavelengths (invisible ultraviolet) and then spit out light with longer wavelengths (visible blue-white light), thanks to fluorescence.

Magnetic Personality

How do I make my own electromagnet?

Go to your workshop or toolbox and retrieve a new D battery, a big iron nail, electrical tape, and several feet of thinly coated copper wire. Wrap the middle of the copper wire around the nail evenly. Leave a good length of the wire free on either end, and strip the coating off the ends. Tape the stripped wires to either end of the battery (i.e., the positive and negative ends).

Point one end of the nail toward metal objects like safety pins or paper clips and watch the attraction. When you've played to your heart's content, disconnect the wires from the battery, but be careful; the electrical charge generates a lot of heat.

How can I construct a hovercraft from a leaf blower that will lift several adults?

Instant Tattoo

Is there any way to make my own temporary tattoo at home?

For a temporary drawing, you can use a marker or henna, of course. But it's even cooler to use a Polaroid picture, and it's an interesting, photorealistic alternative to gumball-machine tattoos. Here's how you do it:

Using a Polaroid camera, take a picture of whatever image you'd like to use—it could be of yourself, your significant other, or something more fanciful and artistic like a bowl of fruit, or the kitchen sink. As the picture shows up, press it against your arm. The developer dye is still wet and will transfer onto your skin.

You can play with these prints in other ways as well. Smear the dye around with a pencil to make abstract art, or take a picture of a friend and, with light strokes, change your friends' hair or clothes, or create a "touched up" look as if the image had been painted.

Which Came First, the Chicken or the Experiment?

My friend had a wishbone that wouldn't break—it just bent like rubber. How did he do that?

Probably by using vinegar, an acid that will dissolve the calcium in bones. Calcium, as you probably know from school or television commercials, is what makes bones hard and strong.

Try it yourself: take the drumstick of a chicken and clean off all of the meat. Place the bone in a jar of vinegar and leave it for a week. Take it out, and you'll have a rubber chicken bone you can use in your vaudeville act.

Where can I buy a rubber chicken? Ask

How do you get a hard-boiled egg to go through a small bottle opening without breaking it?

This is a fun trick that sounds even more impressive before you see it than after.

That's because you use a peeled egg. If you want to add a little more scientific mystery, you can combine this with the rubberized-chicken-bone stunt discussed above. It turns out that if you soak a hard-boiled egg in vinegar for several days, the shell is eaten away, leaving the egg wrapped only in that rubbery membrane that lines the shell.

Begin with a bottle that has a mouth just a little smaller than the egg (for example, an old-fashioned milk bottle, or a big ketchup bottle). If it's too big, the egg will fall through before you begin; if too small, the trick won't work.

Boil some water. Now get ready to work fast. Pour the very hot water into the bottle using an oven mitt. Swish the water around until the bottle is quite hot, and then dump it down the sink. Quickly balance the egg in the opening of the bottle and wait for a few seconds. Drum roll, please—as you watch, the egg will be sucked into the bottle.

So how does it work? The boiling water heats up the air inside the bottle, making it expand. When the bottle begins to cool, the air contracts. Since the air pressure outside is now higher than the air pressure inside, it pushes the egg into the bottle.

Okay, so now you have an egg inside a bottle. Now's the time to challenge your audience with the question: How do you get the egg out without breaking it? It's possible that reversing the process will work—swishing ice water around inside the bottle and dumping it out before turning the bottle upside down should theoretically work. However, there's an easier way. Turn the bottle upside down and blow into it. The egg will move out of the way and allow air in. The increased air pressure inside the bottle should then cause the egg to pop out.

Can you tell me when is the best time during the spring equinox to get an egg to stand on end?

Any time at the equinox is a good time to try. Or, for that matter, on Easter, Hanukkah, Arbor Day, Ramadan, Shrove Tuesday, or your mom's birthday. The truth is that if you're patient or steady enough, you can get an egg to stand on end at any time, on any day of the year.

The egg-standing spring equinox urban myth is one that just won't die. The idea is that "the sun's gravity lines up with Earth" on that day, or some other such foolishness. It isn't true. (If you're interested, writer and myth debunker Martin Gardner actually tracked down the origins of this one. An American writer in China reported in *Life* magazine about a custom in a Chinese village of balancing eggs "on the first day of spring." When other magazines and papers picked it up, they left out an important fact—that what the Chinese consider "the first day of spring" actually takes place in early February.)

Here's your homework: go get a dozen fresh eggs from the grocery store today—DON'T wait until the equinox, or you won't believe us. When you get home, find a very sturdy, flat surface. A good, flat kitchen counter works well, but if you have access to a lab counter, all the better. You want something that won't shake so the eggs won't tumble over.

Now practice standing the egg up. If you're really, really patient, you can get an egg to stand on end without any outside aid. If you're impatient, you can shake the egg, breaking the yolk away from the albumen (egg white) cords inside, which some egg balancers say allows it to settle a little deeper. Another egg balancer suggests that letting the egg warm to room temperature makes it balance more easily, possibly by reducing condensation, which makes the egg slippery.

Why can an egg stand on end? It turns out that an egg isn't as smooth as it seems. It has hundreds of little pores that give the egg a little surface irregularity to rest on. In fact, if you want to cheat, you can also sprinkle salt on your surface to give the egg more little bumps to balance with.

If you really want a challenge, practice balancing it on its smaller end. It is possible, but harder.

Slime after Slime

How can I make slimy, gooey stuff like Basic Fun's Slime, for my slimy, gooey kids?

First get white glue, borax, food coloring, and water. Take a cup of water and add one tablespoon of borax to it and stir. Take a half cup of water and add a half cup of white glue. Add the two mixes together with some food coloring, mixing thoroughly (an easy, not too messy way to do this kneading is inside one of those zip-locking food-storage bags).

For an even grosser variation, mix in your favorite repugnant things—rubber maggots, eyeballs, flies, etc. Store it in an airtight container to keep it fresh and slimy, and do what you can to keep it out of hair, carpets, furniture, and pets.

Sweet Sugar in the Rock

How do I make rock candy?

Homemade rock candy is easy, fun, and illustrates the scientific concept of crystallization. It also teaches patience, because you have to wait a long time before you can actually eat the stuff.

First get a jar, a pencil, and some twine. Tie the string to the pencil and prop the pencil over the mouth of the jar so the string hangs down into the jar, reaching the bottom without bunching up. You'll want to attach a nontoxic weight to the bottom of the string (say, a stainless-steel paper clip) so that it hangs straight down. Heck, since you're going to the trouble, you might want to prepare several jars and double or triple the recipe below.

In a saucepan, bring a cup of water to a boil. Slowly stir in two cups of sugar, adding half a cup (or less) at a time and waiting for it to dissolve completely before adding the next batch. When all of the sugar has been dissolved into a thick, clear syrup, pour the mixture into your empty jar. Make sure the string is submerged into the hot liquid. Put the jars in a place where they won't be disturbed or tipped over.

Because the water mixture was supersaturated with sugar, the two couldn't stay mixed except when very hot. As the mixture cools, the sugar will begin to glom onto the string, creating cool crystal patterns. As the water evaporates further over the coming weeks, the crystals will grow bigger and bigger.

If you want to play a sadistic practical joke on someone, you can surreptitiously make a special batch of similar-looking "rock candy" by using salt instead of sugar.

Crystal Blue Persuasion

My grandmother used to talk about her "Depression flower garden." What is that?

A Depression flower garden is also called a coal garden or a crystal garden. It's a homemade set of colorful crystals. They're a lot of fun for kids and grown-ups, and are a hands-on way of learning about crystal formations.

You begin by breaking up several charcoal briquettes and putting them into the bottom of a large glass dish. In a cup, mix three tablespoons of salt and three tablespoons of water. Stir well. Add to the salt mixture three tablespoons of ammonia and three tablespoons of bluing (found in the laundry section of your grocery store). Stir well. Be very careful not to get your face too close or spill this on your clothes as you prepare it. If any gets on your skin, wash thoroughly.

When your mixture is ready, carefully pour it over the charcoal briquettes you've layered in the bowl. Sprinkle it with some food coloring—your choice of colors—or you can leave it alone and let the natural colors alone make the magic. Cover the bowl tightly and let it sit for several days, after which you should see crystals forming. If you'd like for them to keep growing, add another round of the chemical mixture of salt, water, ammonia, and bluing, taking care not to pour the liquid directly onto the crystals that have already formed.

There's no real chemical reaction taking place in a crystal garden. What's at work is evaporation. The ammonia speeds up the evaporation of the liquid at the bottom of the bowl. The charcoal is a conduit for the salt and bluing particles (an iron powder called ferric hexacyanoferrate) to climb up when enough of the liquid has evaporated that they can no longer be supported in the mixture. Adding more solution to your dish layers more particles on top of the old formations, making unique and intriguing shapes in a variety of colors.

When Volcanos Aren't Enough Anymore

What's a cool trick to do with baking soda and vinegar besides the tired old papier-mâché volcano?

Baking soda and vinegar probably make the best-known kitchen science combination around. There are few that have results quite as impressive.

Here's an alternative to the age-old volcano. You will need a leak-proof Ziploc sandwich bag, vinegar, warm water, a tissue, and two tablespoons of baking soda. Now place about a half cup of vinegar and a quarter cup of warm water into your Ziploc bag. Carefully zip the bag halfway closed.

Pour the baking soda into the tissue and twist the top. This will prevent the baking soda from spilling into the mixture immediately when it's added. Insert the baking soda tissue into the bag, but don't let it fall into the liquid yet. Instead, hold onto it from the outside while you completely seal the baggie. Let the tissue fall into the liquid and run. When the baking soda, which is a base, hits the vinegar—an acid—the mixture reacts, creating carbon dioxide. The carbon dioxide fills the Ziploc bag pretty quickly, causing it to burst open with a bang. Oh, wait, did we forget to mention that this experiment is best done in a bathtub or outside? Sorry!

Here's another, quieter, less messy, and therefore less interesting way of causing the same reaction. Get a soda bottle and pour a half inch of water and a half inch of distilled vinegar into it. Now take a deflated balloon and see if you can get some baking soda into it. Fit the end of the balloon over the mouth of the bottle without letting the baking soda spill into the liquid yet.

Hold the balloon securely on the bottle and let the baking soda drop down into the liquid mixture. The carbon dioxide that's released will blow up your balloon.

Mad Scientist Baby-sitter

I need a trick I can use to entertain my overactive seven-year-old while we're waiting for our food at restaurants. Any ideas?

1. Start with a glass of ice water. Pull out a thread or a strand of hair. Start by asking your seven-year-old if he can use the thread to lift an ice cube out of the glass. Let him work on it awhile until he gives up, then take the thread back.

   Lay the thread or hair across the top of a piece of ice. Using the salt shaker, shake a little bit of salt across the thread and on top of the ice cube. Wait for a few seconds, make a gesture—presto, change-o!—and gently pull on the thread. The ice cube should attach itself to the thread and come up at the same time.

   Here's what's happening: Salt has the ability to melt ice. The cold of the ice, though, will refreeze the briny mixture as the salt water gets diluted, freezing the piece of thread (or hair) onto the ice cube.

2. Next, try making a straw duck call. Take a straw and smash one end until it's flat. Using a pocketknife or scissors, cut the flattened end into a V-shaped point. Blow hard into the flattened end; you'll get a sound like a bird call or an oboe. Try this with varying lengths of straw, and you'll find that you can play different tones.

   What's happening is the flattened end is vibrating—like all sounds do—at a speed fast enough to make a wacky sound, but slow enough for our ears to hear the vibrations. When the waiter comes to ask you to keep it down, we'd suggest you move on to the next trick.

3. All you need for this one is a clear glass of carbonated water or clear soft drink like Sprite or 7-Up and a small box of raisins. Drop your raisins into the glass of carbonated liquid. Sit back and enjoy the show.

   As you probably know, carbonated drinks are made bubbly by adding carbon dioxide gas to the liquid while it's under pressure. The surface area of the raisins is so crinkled, it'll trap the carbon dioxide bubbles, making the raisins buoyant. As the bubbles form and pop, the raisins will float, bob, and dance.